Upheaval from the Abyss

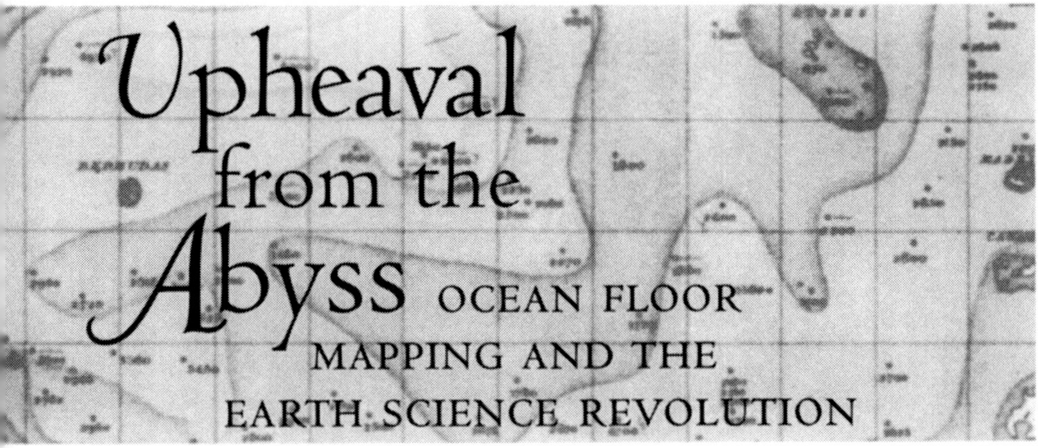

Upheaval from the Abyss

OCEAN FLOOR MAPPING AND THE EARTH SCIENCE REVOLUTION

David M. Lawrence

Rutgers University Press
NEW BRUNSWICK, NEW JERSEY, AND LONDON

Library of Congress Cataloging-in-Publication Data

Lawrence, David M., 1961–
Upheaval from the abyss : ocean floor mapping and the earth science revolution / David M. Lawrence.
p. cm.
Includes bibliographical references and index.
ISBN 0-8135-3028-8 (alk. paper)
1. Ocean bottom. 2. Geological mapping. I. Title.
GC87 .L38 2002
551.46'084'0223—dc21
2001031789

British Cataloging-in-Publication data for this book is available from the British Library.

Manufactured in the United States of America

To
Roosevelt "Critter" Crosby,
who taught me about science;
Laurence Hardy,
who taught me to be a scientist
and introduced me to plate tectonics;
Lonny Lippsett,
who assured me that a sportswriter could handle life
in the "news" world
and suggested that I write about Marie Tharp;
and Sam Freedman,
who infused me with the skills and confidence
to write a book

CONTENTS

PREFACE AND ACKNOWLEDGMENTS

I was first introduced to continental drift and plate tectonics in the fall of 1981 as an undergraduate student in Laurence Hardy's biogeography course at Louisiana State University in Shreveport. Hardy was hardly a dynamic teacher, but he was always well prepared. If you failed to learn from him, you were probably asleep, as I usually was in a class that began at 8:00 A.M.—or in any other class, for that matter. Despite preparation of my own—finding a seat next to a wall so that I could have something to rest my unconscious head against—I somehow managed to learn a thing or two.

One thing that I noticed—probably only by glancing at the syllabus—is that Hardy devoted at least half the course to plate tectonics. He was relentless in his lectures, going on and on about how the concepts of plate tectonics were developed and proved. It seemed odd for a man who devoted his life to studying salamanders, lizards, and snakes and collecting roadkill on dark, rainy nights to spend so much time talking about geology, but Hardy, trained in the fine art of classifying life, was fascinated by the variety of animals and plants on earth. Biogeography, like Charles Darwin's theory of evolution, is vital to a proper understanding of how that diversity arose. And biogeography cannot be truly understood without plate tectonics.

Hardy's enthusiasm infiltrated my somnolent state and infected me. As a result of his class, I embarked on a scientific career that embodied aspects of both biology and geography. And I have maintained an interest in the theory of plate tectonics and its scientific ancestor, continental drift.

In January 1995 I began working as a research assistant at Lamont-Doherty Earth Observatory. Lamont-Doherty has enjoyed a storied history in its relatively young but influential life among the world's great research institutions. As I walked across the observatory's somewhat idyllic campus on the Palisades of the Hudson River, I realized that I was at the place where many of the events Hardy had talked about in his biogeography class occurred. I still feel a rush of excitement to think I have some connection, however tenuous, with that great era of discovery.

Later that year, after two previous brushes with journalism—both as a sports correspondent—I felt that maybe I had missed a calling. I had fallen in love with the chaos of the newsroom. I still thoroughly enjoyed research but longed for a career where I could communicate the mysteries of science to the public. I certainly knew more than most alleged science reporters about the topic—and I also knew how to write a reasonably well-crafted story in an unimaginably short span of time.

Lonny Lippsett, a science writer for Lamont and graduate of the Columbia University Graduate School of Journalism, patiently listened to my inquiries about the science-writing life, and was generous with advice and encouragement. He also suggested that I would find Columbia's journalism program worthwhile and encouraged me to apply. Shortly after my admission into the journalism school, when my relationship with my supervisors at Lamont deteriorated and my career prospects there seemed terminal, he assured me that even a lowly former sportswriter could handle the prestigious work of a "news" desk.

Lonny proved to be right on both counts.

I began attending Columbia University as a journalism student in the summer of 1996. I was in the science/environment journalism concentration, which meant that I was required to take a six-hour seminar on the topic during the spring. I was, however, interested in taking a book-writing seminar taught by a man I was in mortal fear of—Samuel G. Freedman—who, when in a state of well-justified fury, manifested himself in a vision so terrifying it could have turned the hair on a young Charlton Heston white.

Sam, despite his occasional frightening incarnations, has the charisma of one of the great prophets. In late November the Columbia journalism faculty and students assembled to discuss offerings for the spring semester. After he spoke, I was determined to overcome my fear and take the class. To get in, however, I had to sell him on a book proposal. Admittance was a competitive affair. Only sixteen students were allowed.

I wasted no time responding. Before I went to bed that night, I had e-mailed Sam a proposal about mapping the ocean floor. For this, Lonny had already prepared me well. On several visits he had suggested that someone should write about Marie Tharp, a former Lamonter who had mapped the ocean floor and, as a result, played a pivotal role in reviving long-dormant discussions about the theory of continental drift—the subject with which I had long been fascinated. After a weekend of negotiation and modification Sam said I was in the class.

That spring I learned that, yes, Sam could be hard—appropriately so—but he proved more supportive than I could have imagined. I became a father shortly after the class began, and nights rendered sleepless by my son, Malcolm Owain, as well as my work at the copydesk of the *Daily Record* in Parsippany, New Jersey, took their toll. My writing during the first few weeks of class was unimpressive at best, but when it became time to work on the proposal that led to this book, I was determined to make sure Sam would not be disappointed by his decision to admit me.

Apparently he wasn't. When Marlie Wasserman, the director of Rutgers University Press, asked Sam if he could recommend any students with a proposal that might be appropriate for Rutgers, he called.

Soon after signing the contract, I began to fully appreciate something that I had been warned about: Writing a book is not easy. The warning seemed rather trite until I had to take up the task myself. Managing my reporting and writing schedule around family and work—and just finding time for myself—proved to be overwhelming at times. I spent a lot of money feeding photocopy machines in libraries so that I could later read source materials whenever and wherever I had the time. I prefer to talk to sources in person, but circumstances and a limited budget forced me to conduct most interviews by phone, e-mail, or fax. Organizing information proved challenging. I had been taught the use of index cards and database software, but more often than not resorted to the "pile nearest the computer" method—and frequently had to cope with tectonic shifts among piles created by the relentless onslaught of two children, a dog, and two cats. Finally, the time came for me to sit down, shut up, and face the computer.

I hope the reader finds the effort worthwhile.

I COULD NOT have written this book without the help of a biblical-scale host of people. Sam and Lonny should probably be at the top of the list: Sam for teaching me that I could write a book and for showing me how best to get the opportunity; and Lonny for being an inexhaustible source of information, criticism, and encouragement.

My sources have been amazingly helpful, many taking time out from busy research schedules to answer my questions and review drafts of chapters. Special mention should go to Marie Tharp and J. Lamar Worzel. Their recollections and memoirs have been invaluable in my efforts to reconstruct events both before and after the founding of Lamont. Frederick Vine has been especially helpful in the discussion of the seafloor-spreading hypothesis, with his recollections of Harry Hess and John Tuzo Wilson, and by educating an American in some of the finer points of English society. James Heirtzler, along with Vine, undertook a daunting task by attempting to help me understand the nature and study of the earth's magnetic field. Heirtzler kept helping me even when he was (reasonably) unhappy by my initial failure to mention the names of many in the Lamont magnetics research group who made major contributions to the success of the plate tectonics revolution. George Hess, son of Harry Hess, allowed me to invade his home while it was under renovation and flip through a number of his father's old files, notebooks, and photographs. Others who have graciously granted interviews include John and Betty Ewing, Jeff Fox, Billy Glass, Dennis Hayes, Xavier Le Pichon, Ronald Mason, Eldridge Moores, Charles Officer, Jack Oliver, W. R. Peltier, Walter Pitman, William Ryan, George Sutton, Lynn Sykes, and Cindy Lee Van Dover.

The book would have been a lot grayer without the images provided by staff at the Danish Polar Center in Copenhagen; Lamont-Doherty Earth Observatory in Palisades, New York; the National Oceanic and Atmospheric Administration Central Library in Silver Spring, Maryland; the Ontario Science Centre in Toronto; and Woods Hole Oceanographic Institution in Woods Hole, Massachusetts. In particular I would like to thank Henning Thing and Leif Vanggaard at the Danish Polar Center, Janice Aithison at Lamont, Carla Wallace of the NOAA Central Library, Valerie Hatten at the Ontario Science Centre, and Joanne Tromp at Woods Hole.

A host of other people answered (or at least tried to answer) my queries about specific details in the book. Among those who helped were Keith Allen; Tiger Avery; Sylvie Bertrand of the Canada Science & Technology Museum; Deborah Day of Scripps Institution of Oceanography; Loretta De Sio of the Office of Naval Research; John Diebold; Sandy Doyle of the Naval Historical Center; John Eastlund; Peter Edwards; William Glen; Regina Haring; Glenn Helm of the Navy Department Library; Christina Holte and Jutta Voss-Diestelkamp of the Alfred Wegener Institute; Cornelia Lüdecke; Lieutenant Robert Mehal of the United States Navy; Gary North of the Geography and Map Division of the Library of Congress; Neil Opdyke; Naomi Oreskes; Charles G. Patterson Jr.; Scott Price of the United States Coast Guard Historian's Office; John Stevenson; Toby Tyrell; Victor Vacquier; John Vandereedt of the National Archives and Records Administration; and the staff of the American Association of Petroleum Geologists Foundation Energy Resources Library. Among this group I would like to single out Patterson, a surveyor from Richmond, Virginia, whose name I found in the Yellow Pages when I was having trouble writing about some of the finer points of his craft. He listened as I read the passage I had been struggling with over the phone, then patiently helped me get to the point where I not only had the passage right but also actually understood what I was writing about.

I could not have accomplished much of my research without the help of the staff of the Library of Virginia, especially the staff of the greatly overworked Interlibrary Loan Department. Special thanks goes to Dave Grabarek, who cheerfully weathered the torrent of requests I bombarded him with and who shared my penchant for butchering the names of Scandinavian and Finnish publications, and Robert Stephens, who put up with my antics when Dave wasn't around and who somehow always managed to laugh at my rather sad jokes.

It would have been difficult for me to write about the experience of being at sea without actually spending some time on the ocean. Dale Miller, formerly of Windjammer Barefoot Cruises of Miami Beach, Florida, arranged for me to spend a week aboard the S/V *Mandalay*, formerly known as the R/V *Vema*—Lamont's legendary flagship and the first to log more than one million research miles at sea—on a cruise from Puerto La Cruz, Venezuela, to Grenada. (I know, it's a tough job. . . .) I would like to thank Dale and the *Mandalay*'s officers and crew, especially Captain Matt Thomas, First Mate Dimas Manrique, Second Mate Angel Orviol, Purser Laura Jo Bleasdale, Activities Mate Jessica Grotfeldt, and Fleet Captain Paul Maskell—who was the first captain of the ship after its refit by Windjammer. I would also like to thank my fellow passengers—most of whom should probably remain anonymous—who made the cruise an enjoyable working vacation.

My fellow classmates in Sam's book-writing seminar—Shira Boss, Josh Brockman, Susan Carney, Yana Dlugy, Bill Egbert,

Naresh Fernandes, Lauren Foster, Michael Goldhaber, Brenda Jones, Patrick Jameson McCloskey, Eva Mortensen, Allison Nazarian, Bernard Stamler, Alex Tarquinio, and Niraj Warikoo—were and are a source of inspiration and encouragement. Thanks, guys. (I'll never forget the greeting you gave me when I returned to class following the birth of my son.) Several other alumni of the book seminar—Tara Bahrampour, Adam Fifield, Mitra Kalita, and Jacob Levenson—have been especially supportive in the time since we've met. (Tara, thanks again for putting me up during that wicked blizzard in 2000!)

Ari Goldman and Carla Baranauckas, who taught me in my Reporting and Writing I class at Columbia, helped me stay motivated as I struggled through my first semester there—and deserve a significant amount of credit for my staying in the journalism school long enough to be able to take the book seminar. Two other Columbia professors, Bruce Porter and Ron Rosenbaum, skillfully guided me through a number of revisions as I prepared the first result of this project, a profile of Marie Tharp, for publication. Cheri Brooks, former senior editor at the magazine *Mercator's World*, published my profile of Marie as the cover story in the November/December 1999 issue of the magazine.

Fellow authors Bill McMichael and Joanne Kimberlin, who used to work with me at the Newport News, Virginia, *Daily Press*, have been generous with their time and advice. My father-in-law, Ian Sinclair, and Rose Ballance, a woman I worked with at the Virginia Department of Environmental Quality, translated some information that I needed from German. Four people read through the entire manuscript and offered valuable suggestions for improvement: my father, George Lawrence, a retired copy editor for *The Times* of Shreveport, Louisiana, and Gannett News Service; Scott Craig, one of my fellow inmates at the copydesk of the *Daily Press*; Bev Orndorff, retired science reporter for the *Richmond* (Va.) *Times-Dispatch*; and an anonymous reviewer for Rutgers University Press. Sam, Lonny, Yana, Jim Cutler, Theo Francis, Henri Grissino-Mayer, Shauna Jamieson, Lila LaHood,

Stacey Moulds, Beverly Picache, James Rosenberg, Mavis Scanlon, and my friends in the First Unitarian Church writers' group also read portions of the book. Despite the long list of helpers, I take full credit for any mistakes and omissions that remain.

During a period when I despaired of seeing any of this project in print, another Columbia classmate, Catherine Aman, gave me reason not to give up.

J. Warren Frazier of John Hawkins & Associates Inc. in New York provided valuable contract advice when I was negotiating with Rutgers University Press. My editor at Rutgers, Helen Hsu, and her assistants, Sarah Blackwood, Suzanne Kellam, and Jill Stuart, have been incredible resources for me; they've been a pleasure to work with as well, as has Will Hively, who copyedited the manuscript for Rutgers.

Pat McMurray, my supervisor at the Virginia Department of Environmental Quality, was very patient and supportive even though I had to repeatedly take time off to interview sources, write, and revise the manuscript. (Thanks, Pat; I know you took some heat for this.)

Finally, I must thank some of my closest friends and family, who have endured too many of my misadventures over the years. Jane Walker was always generous with class notes when I zonked out in class at The University of Virginia and has been generous with much, much more since. Bill Martin, along with a couple of characters named Bill and Bob, have helped me keep my head together while I learned how to live my dreams rather than just dream them. Steve Stephenson, an unrepentant punster, and Steve Adams, his weary straight man, have kept me mindful that there's more to life than writing and work—there's ecology research, and I should plan to make a trip to the mountains whenever the weather is right.

My father and my mother, Kathy Lawrence, have seen me through disasters too numerous, and occasionally humorous, to mention—it's nice to have something to show for *their* efforts. My cousin (actually my dad's first cousin) Lorraine Persch and her husband, Tom, have helped keep me laughing. Lorraine and Tom also help me remember that I'm basically a swamp rat at heart. (Now if Lorraine could remember Eddie's gumbo recipe, that would be perfect. . . .) And last, here on paper but not in my life, I must thank my wife, Alison Sinclair, our son, Malcolm Owain, our daughter, Mei Seul, and our four-legged critter contingent, Curly, Bali, and Jinn, for putting up with long hours of neglect while I sat in front of the computer. Malcolm, you can play Frogger as soon as I'm finished. . . .

Upheaval from the Abyss

The revolution in thought, if the theory is substantiated, may be expected to resemble the change in astronomical ideas at the time of Copernicus.

—*an anonymous reviewer on Alfred Wegener's theory of continental drift*

. . . utter, damned rot!

—*William Berryman Scott, former president of the American Philosophical Society, on Alfred Wegener's theory of continental drift*

It is impossible to be a geologist without realizing that—in the dim light of the knowledge we have so far gained—the earth we live on is a strange and most improbable planet.

—*Arthur Holmes*

Tales of Mystery and Imagination—and Hard Work

The researcher's life, as I and many others have lived it, is usually characterized by long periods of repetitive, mind-numbing drudgery broken (we hope) by brief explosions of revolutionary insight and frenetic activity. It is similar to the life of a soldier at war, and—at times—can be just as fatal to the participants.

Over the years I have had to battle weather, terrain, dangerous or annoying animals and plants, exhaustion, hunger, stupid yet potentially deadly mistakes, and difficult colleagues in order to collect the data on which my discipline survives. Many of my colleagues have spent years wandering through forests, scrambling up and down mountainsides, peering through microscopes, and poring through stacks of computer printouts in the effort to discern a few secrets of our planet's past. Some in my research community have died, too, from accidents or from ailments brought on by obsessive work.

Historians of science typically analyze the relatively rare bursts of scientific insight. They are justly captivated with the genesis of ideas, the significance of data, and the progress of scientific debates, but they often dismiss drudgery as unimportant and danger as too scandalous to mention during polite, academic discourse.

I believe it is as important to remember the day-to-day human experience as it is to remember ideas, data, and debate. Knowing about the drudgery and danger adds to, rather than detracts from, the history of science. Thus, I have striven to re-create what it is like to be on or under the ocean surface, combating cantankerous equipment or furious weather in the hope of extracting a few precious observations from the deep. I have tried to reveal how some researchers struggled for years inside small labs to digest the increasing flood of information about Neptune's realm, and how others harvested the tide to revolutionize our understanding of the earth's history.

The narrative covers the development of the theory of plate tectonics from its genesis as an often-dismissed theory called continental drift to its triumphal acceptance in the latter half of the 1960s. After years of neglect—even ridicule—the "overnight" success of the idea that massive segments of the earth's crust move has been justifiably called a revolution. The radical change in understanding forced upon the earth sciences ranks with the intellectual upheavals caused by the ideas of Copernicus, Newton, Darwin, and Einstein.

Revolutions are launched by visionaries, and that of plate tectonics is no exception. Alfred Wegener—a German astronomer-meteorologist-physicist-geophysicist-geographer (he was called many things during and after his lifetime, and the monikers were not always complimentary)—was the first to comprehend that proof of the concept that earth's crust was mobile required evidence from many scientific disciplines and that, once proved, the concept answered questions in a diverse range of scientific fields. Wegener was also courageous. He never shirked what Sir Ernest Shackleton called the "white warfare" of the polar regions or the "red warfare of the fields of France and Flanders," nor did he avoid risking the professionally murderous ridicule of his colleagues. His theory of continental drift, which was not well received in his lifetime, laid the foundation for what earth scientists now call plate tectonics.

The story begins in a place that may strike readers as odd, more than a mile above sea level on top of the great Greenland ice sheet, and chronicles Wegener's final expedition to the northern island. This particular expedition had no effect on the development of his theory of continental drift, but it reveals much about his character. His ability to keep fighting despite overwhelming odds, I believe, was one of the primary reasons why his ideas survived decades of abuse before others used them to create a unified vision of how features on earth's surface are created and destroyed.

After recounting the genesis of and early reaction to Wegener's theory of continental drift, the narrative moves to the sea, beginning with early efforts in the nineteenth century to learn about the nature of the ocean floor. The contributions of technologies arising from the ashes of disaster and war are likewise chronicled, but the primary focus throughout remains on the humans engaged in the research effort.

This story contains most of the classic conflicts of great fiction—humans against nature, humans against their fellows, and humans against themselves. It is also a love story—love of the sea, of science, of competition with one's colleagues, and of one another. I hope my effort succeeds in rescuing this tale from the orderly, sterile environment of academic history and giving it the vigorous, combative, lusty humanity it deserves.

NOTE TO READERS: Some may be confused by my use of the term *ocean floor mapping* in the subtitle to this book. Many different types of data fueled the plate tectonics revolution—more than the information about elevation, or depth, most people think of when discussing ocean maps. However, few of the findings covered in this narrative made sense until the data collected were placed in a spatial context—that is, mapped.

In an effort to make this narrative more palatable to general readers, I have chosen not to use endnotes or footnotes in the text. My sources are listed in the "Notes on Reporting" and "Selected Bibliography" sections of the book.

Death on a Glacier

The motor sledges failed. Most of the Greenlanders turned back, taking twelve dog teams with them. The furious winds and temperatures colder than fifty degrees below zero heralded the oncoming winter night and battered the remnants of what was supposed to be a relief mission. But Alfred Lothar Wegener and two other men pushed on to rescue two fellow members of the expedition who were stationed at a research outpost called Eismitte, or Midice, 250 miles inland at the center of the funereal whiteness of the Greenland ice cap.

Wegener—a German scientist and Arctic explorer who had conducted groundbreaking studies in astronomy, climatology, and geology—had spent years overcoming trouble in the pursuit of his diverse interests. The biggest problem for his current expedition of 1930 had been the weather. Sea ice had kept the expedition offshore for more than a month, delaying the establishment of a base camp on top of the icy plateau at the head of Kamarujuk Fjord. Technical difficulties, too, interfered with Wegener's plans. The two motor sledges that were to deliver supplies to Eismitte could not reach the outpost as the engines struggled in bitter autumn temperatures and the skids bogged down in the deep, soft, drifting snow farther inland.

Now all depended on the three men and three teams of dogs.

Most of the time, the animals were virtually swimming in snow up to their bellies. Ice, when encountered, was worse, cutting the dogs' feet as the animals pulled their heavy loads inland.

The slope worked constantly against the group as it fought its way from an elevation of 3,200 feet at the base camp above Kamarujuk Fjord to 9,000 feet at Eismitte. The conditions the group faced also grew more hostile. Temperatures dropped further and further below zero, and the wind blew faster and from the east, working with the slope to impede the party's progress. Snow crystals shot through the air as if fired from a heavenly sandblaster, attacking any exposed flesh.

While the sun was up, the men had to protect themselves from snow blindness, which could prove disastrous to the fate of such a small group. During twilight and darkness, however, the men struggled to find the flags in the snow that marked the path to the distant outpost.

Toward the end of the journey, the men were always cold, even in the shelter of a tent with a Primus stove lit. Their clothes, soaked with snowmelt and sweat, never dried. Ordinarily simple tasks, such as cooking or pitching the tent, proved challenging. More difficult tasks, such as untangling the dogs' traces, became ordeals. Mornings were the most difficult, as the men forced themselves to put on their gear and face even worse cold outside. Yet concern for the fate of their compatriots at Eismitte drove them on.

WEGENER, seasoned by two extended Greenland campaigns, well knew how deadly, fickle, and vast the icebound wastes could be. He first came to the island with the Danmark Expedition of 1906 to 1908. Ludwig Mylius-Erichsen, the leader, and two others were killed by low temperatures and starvation as they tried to return from a trek along the northeast coast in the summer of 1907. Wegener again set foot on Greenland's shores in September 1912 as a member of a team led by Johan Peter Koch, a fellow veteran of the Mylius-Erichsen expedition. A number of members of the team, including Wegener, were nearly killed by falling ice while they ascended the glacier. Two months after their arrival, Koch fell fifty feet into a crevasse and broke his leg.

Koch recovered, and in April 1913 he, Wegener, Vigfus Sigurdsson, Lars Larsen, five Icelandic ponies, and a dog set off on a 700-mile crossing of the ice cap. Gales and blizzards halted the men and animals as they ascended the nearly 10,000-foot-high expanse of ice. Because of the delays, food ran short. The ponies suffered from snow blindness, succumbed to exhaustion, and had to be killed. The men reached the west coast on July 4 but were miles away from the nearest settlement. They rested, cached what they couldn't carry, and made for Prøven, their planned destination. A severe storm pinned the men down for thirty-six hours. Finally, the weakened men, facing starvation, sacrificed the dog. As they were about to eat, the four men were rescued by Greenlanders and returned to civilization.

Wegener had accomplished much during his first two Greenland journeys. He established a reputation as a polar explorer and pioneered the use of balloons and kites to study the atmosphere over the ice cap. While not exploring Arctic land, he published well-received monographs on the thermodynamics of the atmosphere, the origin of lunar craters, and the climate history of the earth. He also published another monograph that, despite three revisions, did not fare as well as his other works. The book was on a topic that later became known as continental drift.

GREENLAND RECLAIMED its hold on Wegener's life during the Easter holidays of 1928 following a visit by Wilhelm Meinardus, a geography professor from the University of Göttingen in Germany, who asked Wegener if he would be interested in leading a small summer expedition to determine the thickness of the ice cap. Wegener knew there was much more research that could, and should, be done. Within weeks he offered a counterproposal to the Notgemeinschaft der Deutschen Wissenschaft, a consortium of German colleges, universities, learned societies, and the Weimar Republic that was to fund the expedition.

Wegener's new proposal added studies of additional ice cap characteristics as well as a battery of meteorological and climatological studies. The most exciting part of the plan, however, was Wegener's suggestion that the expedition overwinter on the ice, not just at one site, but at three along the seventy-first parallel: one each at the eastern and western edges of the glacial massif and one in the middle—Eismitte. The plan would allow continuous monitoring of climatic conditions in the center of the ice cap for a year. The observations in the center, when combined with similar data from the eastern and western edges, would answer a number of questions about the development and passage of weather systems over Greenland. The consortium endorsed Wegener's plan.

Wegener, accompanied by Johannes Georgi, Ernst Sorge, and Fritz Loewe, scouted the western coast of Greenland in 1929 to locate the best base from which to ascend the ice cap and to test equipment and techniques for measuring the thickness of the ice. Probing along Kamarujuk Fjord in a motorboat named the *Krabbe,* they found a suitable candidate: Kamarujuk Glacier, a half-mile to mile-wide tongue of ice bounded by steep valleys that led to the edge of the ice cap more than 3,000 feet above sea level. Wegener decided to establish the western station at the base of a nearby peak, Scheideck Nunatak. As the waters off the west coast of Greenland had the longest ice-free period, Eismitte, 250 miles inland, would have to be established and supplied from Scheideck.

Wegener worked feverishly throughout the winter of 1929 to organize and supply the expedition. In addition to himself, Georgi, Sorge, and Loewe, he selected seventeen other men for the journey. Three were to work independently at the eastern station, which was to be based in Scoresby Sound. Georgi, Sorge, and Manfred Kraus were to overwinter at Eismitte; the rest, including Wegener, were to be based at Scheideck to conduct their own research and to support the trio at Eismitte.

On April 1, 1930, Wegener, with all but three members of the expedition and with enough supplies to fill nearly ten railcars, steamed from Copenhagen, Denmark, on board the *Disko,* the biggest ship that traded with Greenland at the time. The *Disko* called at Reykjavík, Iceland, to pick up three Icelanders, including Vigfus Sigurdsson, and twenty-five Icelandic ponies, and arrived at Holsteinsborg, on the west coast of Greenland, on the fifteenth.

The *Disko* was not built to withstand the icy conditions that prevailed farther north. The *Gustav Holm,* fitted with ice sheathing and a crow's nest to aid maneuvers through pack ice, met the expedition in Holsteinsborg four days later. The cargo, men, and horses were transferred to the *Gustav Holm,* and the expedition began steaming north on April 27.

The *Gustav Holm* was tightly packed, and the cargo, laden with such items as gasoline, dynamite, detonators, and hay, caused considerable concern.

"We have a d——d risky cargo on board," Wegener wrote in his diary entry of April 30. "If fire breaks out we're done for; no hope of putting out petrol. The only consolation is that we shall have a very imposing and expensive cremation ceremony."

When the *Gustav Holm* neared the entrance to Kamarujuk Fjord on May 4, Wegener and his men found that ice blocked their path. The *Gustav Holm* anchored off the nearby Kekertat Islands, and the men, using the ponies and dog-sled teams, began unloading the ship at the edge of the ice and setting up a depot at Uvkusigsat, a village on Greenland's shore six miles away. By the tenth they had finished unloading and were given a hearty send-off by the *Gustav Holm* before it steamed away.

Wegener and his men were all but stranded at Uvkusigsat. The ice was too thick to allow the motorboat *Krabbe,* which had joined the expedition on April 29, to get through to Kamarujuk Fjord, yet it was too rotten in areas to trust with the weight of ponies or heavily laden sledges. They could reach the base of Kamarujuk Glacier using a combination of lightly packed sledges and collapsible boats, but could not move any heavy equipment—including huts and the motor sledges—until the ice broke up. Sometimes the men would try to blast their way through with dynamite, but the ice and weather would not cooperate. Mostly the men waited.

"Whit-Monday, thirty-first day of waiting," Wegener wrote from a camp at Kamarujuk on June 9. "Weather gloomy and my mood ditto. From our view-point on the moraine we can still see the ice fast in Ingnerit Fjord. The programme of the expedition is getting seriously endangered by the refusal of the ice to move. Time is slipping by, and anything we can do here in the absence of ponies and baggage really amounts to very little."

Wegener and his men were stymied for thirty-eight critical days, until June 17, before they could reach the narrow, stony beach at the foot of the glacier in the *Krabbe* and a schooner, *Hvidfisken,* and set up their base.

The men then unloaded 2,500 boxes, cases, and cans. They pitched their tents in the shelter of a moraine and began to organize their supplies. To get the motor sledges to the top they would have to build a road on the glacier. In the meantime they used ponies and dog teams to begin hauling supplies in preparation for the first big journey inland.

Wegener's presence was required at Kamarujuk while the western station was being set up. Thus he could not accompany Georgi, who with Loewe, Karl Weiken, ten Greenlanders, and twelve dog teams set off on July 15 to set up Eismitte. Difficulties arose almost immediately. The next day most of the Greenlanders, who dreaded traveling on the ice cap, thought the load was too heavy for the deep, soft snow they encountered. Georgi cached three boxes of equipment, weighing three hundred pounds, and two hundred pounds of stores. Even with the reduced weights the Greenlanders were reluctant to go along.

According to plan, two of the Greenlanders turned back toward Scheideck on July 17. The rest of the Greenlanders were ready to turn back again the next day because of a misunderstanding over the distribution of rations.

Georgi's party reached the 62½-mile mark on July 19. As they went along they marked the route by erecting snow cairns with route markers every five kilometers (about three miles) and black flags on three-foot-long staves every 500 meters (about 550 yards). On the twenty-second they reached 125 miles. At this point the party was to split up. Loewe was to return with four Greenlanders while Georgi, Weiken, and the four remaining Greenlanders were to push on. But crisis erupted again as the Greenlanders as a body refused to go farther, claiming they "would be unable to breathe, the dogs would die, and at the end they themselves would have to march on foot and eat their boot-soles to still their hunger."

After several hours of tense negotiations—hindered by language differences—the Greenlanders relented when Loewe threatened to go on to Eismitte with Georgi and Weiken and leave the Greenlanders to find their way back to the base camp at Scheideck on their own. At the Greenlanders' insistence, Georgi further reduced the load of supplies slated for the outpost. Then he, Weiken, and the rest of the 250-mile party forged on. At six in the evening on July 30 the men reached their destination. The next day they put up the tent and hut, unpacked the supplies, and assembled some of Georgi's scientific instruments. Georgi wrote letters to his family and to Wegener, giving them to Weiken to take back to Scheideck. Weiken and the Greenlanders left Georgi alone at Eismitte on August 1.

If all had gone according to plan, Eismitte would have been resupplied by a combination of three dog-sled trips and one or two motor-sledge trips. Loewe reached Eismitte with a second dog-sled party on August 18. Sorge, accompanied by Kurt Wölcken, Hugo Jülg, seven Greenlanders, and ten dog-sled teams, joined Georgi on September 12. But the motor sledges never appeared.

The two motor sledges were specially built for the Wegener expedition by Finland's state aircraft factory in Helsinki. The sledges, christened the *Schneespatz* and *Eisbär*, looked something like toads, each with a huge propeller on the back (as on airboats used in the Florida Everglades) and each mounted on four broad skids of hickory with rubber springs. The front skids could be turned like the wheels of a car. Each sledge, containing a cabin for the driver and plenty of storage space, was powered by a 110-horsepower Siemens aircraft engine and carried a sixty-three-gallon fuel tank.

The Finns used motor sledges with great success to travel between islands off their coast in winter. But travel at sea level was different than at several thousand feet above. Wegener's sledges were often buried in snow or frozen in during inclement weather. If the skids were not encased in ice and snow, then the propellers or engines frequently froze up. Even when the engines were not frozen, they did not run as efficiently as at sea level, because of the lower oxygen content at high elevations. Fuel consumption also increased in the deep, soft snows that fell on the ice cap, especially if the drivers had a difficult time following the trail.

Regardless, Curt Schif and Georg Lissey in the *Schneespatz,* and Kraus and Franz Kelbl in the *Eisbär,* set off from the 53-mile depot for Eismitte on September 17. Much of their load and additional fuel had already been cached at the 125-mile depot, which they reached in five hours. There they met the Wölcken and Jülg party returning from Eismitte. The next morning the motor-sledge crews were beset by thick mist and driving snow. While the dog teams, which could travel in virtually any weather, left for Scheideck, the motor-sledge party had to wait. They could ill afford to get lost under those conditions.

On that day and the next, the sledges were nearly snowed in. When the weather improved on the twentieth, the *Eisbär*'s engine started with difficulty, while the *Schneespatz*'s engine would not budge. Eventually it yielded—after being heated for an hour and a half by a Primus stove and soldering lamps—but the men were too exhausted to do anything else after hours of struggle at 8,200 feet. The weather was dismal again the next day. On the twenty-second, the weather improved, but heavy drifts and a fierce headwind prevented the sledges from making any progress toward the east. There was no way the motor sledges could make Eismitte with a load. Thus, only a day away from their destination and running out of their own supplies, the motor-sledge crews turned back.

Among the casualties of the failed motor sledge trip were a winter hut Georgi and Sorge were to live in, more food, extra kerosene for fuel, and a wireless for communication with Scheideck. Georgi and Sorge discovered they could survive the winter in an ice castle built from the snow. After some calculations, they determined that their food and kerosene supplies would last them until the following spring, when they expected to be relieved. The least important item for their immediate survival was the wireless. But the lack of the wireless was to play a crucial role in the events that followed.

If all had gone according to plan, either the motor sledges or a fourth visit by dog-sled teams should have reached Eismitte by September 20. Georgi and Sorge initially had planned to leave and make their way back to Scheideck by man-hauled sleds if no further relief came by October 20. Without the wireless the pair could not contact Wegener and tell him that they had enough supplies to survive the winter and had decided to stay.

Late in August Wegener realized another dog-sled journey would probably be needed to fully provision Eismitte. On September 4 he asked Weiken and Loewe to begin planning to take a fifteen-sled trip. Deeply worried about the welfare of Georgi and Sorge, however, he decided on the eighteenth to go on the journey himself and left Weiken in charge of the expedition at Scheideck.

Three days later Wegener, Loewe, and thirteen Greenlanders departed Scheideck, meeting the returning Wölcken and Jülg party only a few miles out. On the twenty-fourth, at the 31-mile mark, they met the motor-sledge party and learned that the sledges had not reached Eismitte.

Wegener didn't need to see the clouds rising in the west that afternoon to be troubled by the lack of progress. He cut the load planned for Eismitte by 1,600 pounds and arranged to send a Greenlander, whose sledge had fallen apart, back to Scheideck on one of the motor sledges. By the next morning the storm arrived.

A raging blizzard kept them confined to their tents for the next two days.

The weather cleared by the morning of the twenty-seventh, but a stiff easterly wind and temperatures dropping to seventeen below zero limited the progress of Wegener's party to just 7½ miles. That night the Greenlanders ominously gathered in one tent. Wegener and Loewe knew something was brewing. The next morning the Greenlanders entered Wegener and Loewe's tent. At first they said nothing, just staring at the ground and smoking their pipes. Finally, one announced that they wanted to turn back.

After a long talk and promise of higher pay, four of the Greenlanders, Detlev Frederiksen and Rasmus Villumsen of Uvkusigsat and Nikola Sakiussen and Johann Amossen of Kekertat, agreed to continue. Wegener gave the others a letter to take back to Weiken, in which he wrote:

> The whole business is a big catastrophe and there is no use in concealing the fact. It is now a matter of life and death. I will not ask you to do anything to ensure our safety on the return journey, for there are plenty of depots. The only help you could render us would be the psychological one of sending a party out to meet us; but in October that would again involve considerable risk to the relief party. I do not consider Sorge's plan of setting out on 20th October with man-hauled sledges feasible; they would not get through but would be frozen to death on the way. — We shall do what we can and we need not yet give up all hope of things going well. But good traveling conditions seem to be definitely over now. Even the journey here was very strenuous, and what lies ahead is certainly not a pleasure trip.

The loss of eight dog-sled teams forced Wegener and his men to cut drastically the amount of supplies for Eismitte. On September 29—nine days after some kind of relief party was scheduled to have reached the outpost—six men, six sleds, and sixty-nine dogs continued on, carrying barely two tons of provisions for the station. Wegener and his greatly reduced party had covered only about 15 percent of the distance to Eismitte. They still had nearly 212 miles to go.

Wegener's party reached the 75-mile cairn on October 1. By now Frederiksen was saying that he was tired and wanted to turn back. Wegener again persuaded him to continue. Another snowstorm pinned them down for a day and forced Wegener and his team to cache the rest of the supplies they had planned to take to Eismitte. Kerosene, the most critical of the supplies needed at the station, could be picked up at the 125-mile depot.

Two days later the party trudged on, but the deep, fresh snow became almost insurmountable, even for the nearly empty sleds. In the summer the dog-sled teams averaged at least 18 miles a day, but between the third and the sixth Wegener's group had covered only 19 miles. At that rate, Wegener knew, there would not be enough food for the men and dogs of all six teams to make it to Eismitte and back.

On the night of October 5, Frederiksen again said that the Greenlanders wanted to return to Scheideck. Fredericksen had good reason to want to return; he had been leading the party along the trail and was exhausted by the effort. Wegener considered giving up the quest, but Loewe convinced him that they could make the station with one of the Greenlanders. At first Sakiussen volunteered to continue, but after everything was repacked and reorganized, he changed his mind and began offering excuses why he should not go on. Finally, on the seventh, Villumsen said he would leave for Eismitte with Wegener and Loewe. The parties split up. Wegener sent another letter back to Weiken at Scheideck. He knew the desperate straits he was in and had changed his mind about a relief party, asking Weiken to send one—consisting of no more than two men and two dog teams—to the 38½-mile depot. Frederiksen, Sakiussen, and Amossen reached Scheideck on the fifteenth. Wegener, Loewe, and Villumsen would not have it so easy.

The weather and deep snowdrifts limited their progress to just 12 miles over the next three days. The flags marking the trail were often buried. Fortunately, Villumsen had a knack for spotting what little of the tips remained visible. The lead dogs sank to their bellies in the snow. The sleds, weighing five hundred pounds even with their reduced loads, frequently became stuck, and the men—often wading knee-deep through the drifts—struggled to get them moving again. Facing temperatures that remained lower than twenty-two degrees below zero, the trio rested on October 10 and reassessed the situation.

Wegener and his men had about two weeks of food left. They tentatively made plans to cache what supplies remained at 143 miles and turn back. But conditions improved on the eleventh. They instead decided to continue until they either reached Eismitte or met Georgi and Sorge on the path.

The party made reasonable progress from then on, but low temperatures and lack of food for the dogs began to take their toll. Loewe began to suffer from frostbite on his hands and feet. The dogs weakened. By October 24 they had made 208 miles and had only 42 miles to go. The next day, however, the temperature never rose above forty below zero, and the men remained in their tents.

The delay was futile. The temperature continued to drop, remaining around fifty-eight below zero. Wegener, Loewe, and Villumsen resumed their journey. The moisture in the breaths of men and dogs froze immediately, enshrouding them in clouds of ice crystals that trailed the party for half a mile. Grasping anything was painful, making an ordeal of the tasks of setting up and striking camp and untangling the dogs' traces. The pemmican for the dogs had to be broken up with an ax.

Loewe noticed that he had no feeling in his toes on the twenty-seventh. Despite hours of massages from Wegener, the circulation could not be restored. The dogs' food supplies ran out the next day. The last of the kerosene was burned by the morning of the thirtieth, but by then they were close. Wegener, Loewe, and Villumsen trudged the last few miles to Eismitte in a temperature of sixty-two degrees below zero and arrived at eleven in the morning. The temperature inside Georgi and Sorge's ice castle was a balmy twenty-three degrees.

There was no way all five men could survive the winter at Eismitte with the meager supplies at hand. Loewe, his feet severely frostbitten, would remain, but Wegener and Villumsen had to face the weather once again in the interest of saving the other men's lives.

"Wegener wants to start back again with Rasmus [Villumsen] very early to-morrow morning," Georgi wrote on October 31 in a letter he sent back with Wegener, "with their dogs pretty well worn out it is a race with death."

On the morning of November 1 the men celebrated Wegener's fiftieth birthday with sandwiches and preparations for his and Villumsen's departure. Photos of the pair were taken before they left. Both men were tightly bundled in reindeer skins and furs so that only parts of their faces were exposed to the elements. Ice had collected on Wegener's mustache, setting it in stark contrast to his grime-darkened skin. The two headed off to the west in a temperature of forty below zero.

Wegener and Villumsen never reached Scheideck. In the spring, relief parties found Wegener's body alongside the trail 118 miles away from the western station. It had been carefully buried by Villumsen, who then left carrying diaries, letters, and other odds and ends. Evidence of Villumsen was found at two campsites, one at 105½ miles and the other at the 96-mile cairn. His trail vanished in the seemingly boundless wastes between the 96-mile cairn and Scheideck.

WEGENER KNEW that, by continuing on to Eismitte as his sled party's supplies ran out, he was endangering his life. But he had never been one to avoid doing his duty. The men who found his body in the grave Villumsen had carefully prepared noted that Wegener had a peaceful, almost smiling, look on his face—the look of a man who died knowing he had done what was right.

The dogged determination to do what he thought was right was evident throughout his career, especially in the defense of his crowning intellectual achievement—the theory of continental drift. For nearly twenty years he had nurtured the theory, even though it had been met with unimaginable hostility within the scientific community. Despite the abuse from his colleagues, he never surrendered. Instead, he constantly improved the theory, bolstered it with additional evidence, and published three revisions to *The Origin of Continents and Oceans*, the monograph in which he presented his ideas. The final edition was published the year before his final expedition to Greenland.

By the time he died, Wegener and continental drift were anathema to most earth scientists, and they remained so for decades. But scientific facts eventually resurrected the debate and in turn resurrected Wegener's reputation. For continental drift proved to be the spark that ignited one of the greatest revolutions in the history of science.

Radical Notions

The spirit of revolution that descended upon the world in the eighteenth and nineteenth centuries and inspired wars of liberation touched off an uprising among the sciences as well. Renaissance thinkers had earlier released the shackles of classical Greek and Roman philosophy that had stifled inquisitiveness and experimentation for hundreds of years. But these were pious times, and most scientists in the Christian world embraced the Bible as the ultimate reference work on the history of the earth and life.

According to the Book of Genesis, the world and all life in it was created over a period of six days. For a work that was supposed to be the final word on historical matters, the chronology of the first chapter in the book proved to be confusing. God, for example, created light and "divided the light from the darkness" on the first day, and called the light "Day" and the darkness "Night." But God created lights in the heavens to "divide the day from the night" again on the fourth day. However, it is clear that the heavens and the earth were created on the first day and plants were added on the third. Beasts of the sea and air were brought forth on the fifth day and told to reproduce. Land animals were created on the sixth day. God added humans, too, on the sixth day and told them to start reproducing. (Plants and land animals must have been enthusiastic enough to reproduce without God's insistence—humans don't seem to need much encouragement today.) In an apparent flashback, chapter two of Genesis, which follows God's rest on the seventh day, tells of the origin of Adam and Eve.

The progeny of Adam and Eve went forth and multiplied, but by the time Noah's generation inherited the earth God had grown weary of human misbehavior and decided to "destroy man whom I have created from the face of the earth; both man, and beast, and the creeping thing, and the fowls of the air; for it repenteth me that I have made them." God's weapon of choice was a massive flood that inundated the earth, and it was reported to be effective. "And all flesh died that moved upon the earth, both of fowl, and of cattle, and of beast, and of every creeping thing that creepeth upon the earth, and every man: All in whose nostrils *was* the breath of life, of all that *was* in the dry *land*, died." (Curiously, fish and other aquatic animals are never mentioned among the list of victims—maybe the recorders of Genesis realized that water creatures would likely survive such a torrent.) Noah and his passengers on the ark survived and repopulated the earth afterward.

This tidy, biblically inspired worldview of scientists began to unravel as the significance of fossils was realized. Many fossils, like those of the dinosaurs, were of organisms that had never been seen in the flesh. Rather than discard the teachings of the Bible, some scholars argued that some fossils represented living organisms that had not yet been discovered, while most others were of species wiped out during Noah's flood. Other observations were more difficult to explain. Leonardo da Vinci was among the first to seriously doubt Noah's flood when he determined that shells found in sedimentary rocks hundreds of miles away from the sea in northern Italy must have had a marine origin. The shells were too far from the ocean and too fragile to have been washed in by the floodwaters, and the duration of the flood—40 days and 40 nights of rain, but 150 days of high water levels—did not give the organisms enough time to migrate to where their fossils were found. Leonardo also took note of the fact that strata containing shells were separated by layers in which no fossils were found, indicating that a single catastrophic event was unlikely to be the cause. Unfortunately, his analysis was recorded in his private notebooks, in his notorious backward script, and had little, if any, impact on science afterward.

Writers of the time went to what now seem bizarre lengths to explain the presence of fossils without discarding their belief in the literal truth of the Bible. Religious authorities claimed that fossils were planted by Satan to lead the faithful away from the truth. Englishman Martin Lister suggested in 1671 that fossils were entirely mineral in origin rather than the remains of ancient life. Another Englishman, Robert Plot, claimed in 1677 that fossils were created by some "plastick virtue" within the earth. Plot neglected to explain how that "virtue" worked. Yet another Englishman, John Woodward, in 1695 proposed an elaborate scheme in which the laws of gravity were suspended during the flood, allowing water and earth to mix into a prodigious muck that settled out in layers as the waters receded.

Finally someone stepped up to free the study of fossils from biblical influence. Robert Hooke—he was English as well—began a meticulous study of fossils in the mid-1600s with a recent invention, the microscope. From his studies, Hooke deduced that individual fossil species had fixed life spans, that some species that had lived in the past were not alive in the present, and that some living in the present had not lived in the past; thus he anticipated the concepts of extinction and evolution. In lectures before the Royal Society of London in 1668 and 1687, he proposed that the earth had a slight wobble on its axis, which caused polar wandering. He said that earthquakes were a result of the buildup and sudden release of stresses within the crust—the release being accompanied by slippage of strata and, possibly, cycles of submergence (and deposition) and uplift (and erosion) of the surface. Hooke also suggested that fossils could be used to develop a chronology of the layers of the surface of the earth.

While Hooke was suggesting that the history of the earth could be revealed through the study of fossils, Nicolaus Steno, a Dane living in Florence, Italy, laid the foundation for such work. While court physician to Grand Duke Ferdinand II of Tuscany, Steno studied the fossils and geology of the region. He published his work in 1669 in a small book, *De solido intra solidum naturaliter contento dissertationis prodromus* (The prodromus of Nicolaus Steno's dissertation concerning a solid body enclosed by process of nature within a solid). Steno outlined three principles critical to the subsequent development of geology. The first, the principle of superposition, states that, in an undisturbed series of sedimentary strata, the oldest layers will be found at the bottom, the youngest at the top. The second principle, that of original horizontality, stipulates that, since sedimentary strata are composed of materials that settle from suspension in a fluid, most undisturbed layers will be horizontal. The third, the principle of original lateral continuity, mandates that sedimentary strata will originally extend horizontally in all directions until reaching the edge of the zone of deposition.

Steno shortly afterward abandoned geology to become a priest, later a bishop, in the Roman Catholic Church. His work, however, had lasting impact.

By the end of the eighteenth century and beginning of the nineteenth, scientists including English surveyor William Smith and two Frenchmen, mineralogist Alexandre Brongniart and zoologist Georges Cuvier, had synthesized the works of Hooke and Steno and realized that, by correlation of fossils from widely separated regions, a chronology of the earth's past could be developed. The results of an increasing number of studies indicated that Hooke was right in that species arose and went extinct at various times throughout the span of geologic history, which began to undermine the idea that all species had been created at once and that a biblical catastrophe was responsible for a mass extinction during Noah's day.

The bomb that finally brought the biblical edifice down was the theory of evolution. The fuse was lit in 1858 when papers outlining the theory of evolution by natural selection, independently developed by both Charles Darwin and Alfred Russel Wallace, were presented to the Linnaean Society of London. The bomb detonated in 1859 when Darwin followed up the papers with his classic *On the Origin of Species by Means of Natural Selection*. The theory of evolution cleared up a number of questions about how species originated and changed through time but also created a host of new questions that would never have come up as long as God was letting things be over a period of six days.

One set of questions concerned the distribution of plants and animals—for example, *Mesosaurus*, a slim, river-dwelling reptile about a yard in length. Mesosaurs are found in deposits about 280 million years old in South Africa and southeastern Brazil. While the reptile was obviously well adapted for freshwater habitats, the famed paleontologist Alfred Sherwood Romer wrote in 1966 that "it is difficult to imagine it breasting the South Atlantic waves for 3,000 miles." If *Mesosaurus* did not swim from one continent to another, why are its fossils found in both?

Even more striking was the distribution of the *Glossopteris* flora. The extensive formation—identified by the presence of the tongue-like leaves of the seed fern that gives the formation its name—is found in deposits between 280 million and 345 million years of age in southern Africa, Madagascar, India, Australia, and South America. The Antarctic explorers Sir Ernest Shackleton and Captain Robert Falcon Scott even brought back fossils from the *Glossopteris* formation from their legendary expeditions to the ice-bound continent (well, in Scott's case the fossils made it back; he and four others perished early in 1912 on their return from the South Pole). Why was this assemblage of plants and animals found in five regions separated by such vast expanses of the Pacific, Indian, and Atlantic Oceans?

Neither question—about the distribution of mesosaurs or of the *Glossopteris* flora—would have arisen so long as the Bible reigned supreme in science, for the answer would clearly be God's will. By jettisoning Genesis for evolution, however, scientists created a problem for themselves: Some means must be provided to allow terrestrial organisms to migrate across thousands of miles of the earth's ocean basins. For much of the nineteenth and twentieth centuries, biologists and geographers postulated long-lost land bridges or sunken continents, such as Atlantis, in the Atlantic Ocean, or Gondwanaland, in the Southern Hemisphere.

AT THE BEGINNING of the twentieth century, scientists struggled to explain the origin of the major features of the earth's surface, such as the ocean basins and the great mountain belts. The genesis of earthquakes and volcanic explosions—which have destroyed, and can still destroy, civilizations—lay hidden deep below the surface in a domain explored only by mythical gods, medieval poets, and fictional scientists. In an era when the urge to collect and classify every living thing imaginable was indulged at the highest strata of society, devising an explanation for the distribution patterns of the plant and animal groups discovered proved as insoluble a problem as counting the angels dancing on the head of a pin. Humans were baptized with the blessings of technology on a grand scale. Almost anything seemed possible. But few earth scientists at the time could comprehend that structures as massive as continents were capable of movement.

Evidence of movement on smaller scales, from the presence of marine fossils in strata high above tree line in the Alps, and from the great folds and overthrusts in the Alps of Europe and Appalachians of North America, was obvious, but explanation, even acceptance, of the obvious took a long time. In the 1780s the Scot James Hutton proposed cycles of mountain building and destruction driven by intrusion of hot magma—which caused uplift of the surrounding terrain—and subsequent erosion of the resultant highlands. While this theory adequately explained the origin of some mountain regions, the idea could not account for the long, linear, parallel features characteristic of many mountain chains.

Sir Isaac Newton as far back as 1681 suggested that mountains had been formed early in the earth's history as it cooled and shrank in diameter. As a result the crust at the surface buckled, thus forming mountains. Elie de Beaumont, a powerful member of the French Academy, warmed to the idea and resurrected it in 1829. Eduard Suess, professor of geology at the University of Vienna, elaborated on the contraction theory in his massive *Das Antlitz der Erde* (The face of the earth), whose four volumes appeared over a period of twenty-one years (1883–1904). Suess compared the concept of a contracting earth to a drying apple. As desiccation proceeded and the surface area of the apple decreased, the skin would wrinkle. Some parts of the skin would be elevated to form mountains; other parts would collapse to form ocean basins. Suess envisioned several cycles of uplift and collapse. The theory thus could explain the origin of the great mountain belts around the earth as well as the distribution of similar flora and fauna on continents now separated by thousands of miles of ocean.

James Dwight Dana, professor of geology at Yale University and longtime editor of the *American Journal of Science* in the latter half of the nineteenth century, envisioned a less dynamic version of the contraction theory. Dana proposed that the differences between continents and ocean basins appeared early on as the earth cooled from a primordial molten state. The continents solidified first; basins opened between the continental blocks, then deepened as the rocks forming the floors of the depressions themselves cooled. In his view, mountains should form primarily at the edges of continents where magma was forced toward the surface by the force of the subsiding ocean basins. Dana's theory seemed to explain some of the great mountain ranges along the margins of the Americas, such as the Appalachians and the Andes. A subsiding ocean floor was in keeping with another of Charles Darwin's great ideas, that of the origin of coral atolls, in which Darwin proposed that coral reefs originally formed along the margins of volcanic islands and continued to grow toward the surface of the ocean as the volcanic base cooled and sank in the depths. Dana's theory, however, was less encompassing than that of Suess in that it could not explain how similar groups of plants and animals could appear in widely separated locations.

One implication of Dana's theory was that the continents and ocean basins were, once formed, permanent features of the earth. The perception of permanence sank into the consciousness of American geologists so thoroughly that Bailey Willis, a Stanford University professor who was later to viciously attack Alfred Wegener's theory of continental drift, wrote in 1910 that "the great ocean basins are permanent features of the earth's surface and they have existed, where they are now, with moderate changes of outline, since the waters first gathered."

AS GEOLOGISTS DEBATED the origin of continents and oceans in the nineteenth century, it had become clear to most that the earth was much, much older than the six thousand years estimated by the venerable Bishop James Ussher in 1650, maybe hundreds and hundreds of millions of years old. But a Scottish physicist, William Thomson (otherwise known as Lord Kelvin), calculated the rate of cooling of the sun and earth and concluded that the earth was too hot, therefore too young, for all the geologists' shrinking and wrinkling to take place. At best, Thomson estimated the sun to be only a few hundreds of millions of years old and the earth only a few tens of millions of years in age. Thomson began his attacks on geological time in 1846 and kept pressing them throughout the rest of the century. The sun and earth got younger as he got older, with Thomson writing near the end of his well-documented eighty-three years that the earth could be no more than twenty (million years, that is). Thus Thomson took part in physicists' long tradition of debunking geologists' pet ideas. Unfortunately for Thomson, he lived up to another hallowed tradition of physicists who try to understand the earth by the use of math alone—his calculations worked just fine, but flawed assumptions undermined his conclusions.

Thomson's age estimates were based on two shaky premises: (1) that both the sun and earth were originally bodies of hot liquid; and (2) that no additional heat source had warmed the bodies since their origin. Filled with the fear of physics, few geologists felt qualified to challenge Thomson on his own terms. But an American geologist, Thomas Chrowder Chamberlin, finally took Thomson on in 1899 in a pair of papers that pointed out that the first assumption was merely an *assumption*, not a fact; therefore it was as subject to question as any other idea in science. Chamberlin, with the foresight of a prophet, added that the discovery of an additional heat source would disprove the second assumption and therefore invalidate Thomson's age estimates. Ernest Rutherford (later Lord Rutherford) and Howard T. Barnes reported their discovery of radioactive heating four years later. George Howard Darwin (Charles Darwin's son) quickly seized upon that idea to propose that a heat-generating substance inside the earth helped keep it warm. The finding, by breaking the Thomsonian stranglehold on the age of the earth, bought geologists time, billions of years, for mountain building and destruction to take place. It also allowed time for larger-scale phenomena, such as the drift of continents, to take place.

SIR FRANCIS BACON has often been credited with being the first to come up with a vision of continental drift. In his 1620 treatise, *Novum Organum*, Bacon wrote:

> The very configuration of the world itself in its greater parts presents conformable instances which are not to be neglected. Take for example Africa and the region of Peru with the continent stretching to the Straits of Magellan, in each of which tracts there are similar isthmuses and similar promontories, which hardly can be by accident.

The statement above has been interpreted by others as an astute observation on how well the landmasses would fit together if joined. Unfortunately, Bacon's "prescience" is a case where truth and a good story have gone their separate ways rather than continents. All Bacon had observed—and commented on—was a similarity in overall shape between the two if *superimposed*, with Peru jutting out like West Africa.

Abraham Ortelius did write about continental drift, however, and he did so in 1596—twenty-four years *before* Bacon's misinterpreted statement. Ortelius advanced his idea in a commentary on Plato's Atlantis in the third edition of Ortelius's famous atlas, *Thesaurus Geographicus*. Plato wrote that the continent of Atlantis had sunk below the ocean in a violent cataclysm. Ortelius—while leaving open the question of whether or not Atlantis existed—suggested that the continent did not sink but had been torn away to the west. Furthermore, Ortelius added that evidence of the rupture could be seen in the new maps of the world:

> But the vestiges of the rupture reveal themselves, if someone brings forward a map of the world and considers carefully the coasts of [Europe, Africa, and the Americas], where they face each other—I mean the projecting parts of Europe and Africa, of course, along with the recesses of America.

Unfortunately, Ortelius's hypothesis passed unnoticed and was forgotten for the next few centuries.

Until the eighteenth century no one besides Ortelius had clearly attempted to explain the coincidence of transatlantic shapes by invoking a rearrangement of components of the earth's crust. The origin of ocean basins was explained by upheavals of the surface during Noah's flood. Theologian Theodor Christoph Lilienthal in 1756 wrote that the continents had been separated as a result of the flood. The great German geographer Alexander von Humboldt suggested several times during the early 1800s that the deluge had gouged out the Atlantic Ocean basin. Antonio Snider-Pellegrini got a little closer to the truth in 1858 by proposing that eruptions of material from the interior of the earth had pushed the continents apart, thus creating ocean basins. Unfortunately, however, Snider also invoked the flood as a cause of the upheaval.

George Darwin, less biblical in inspiration but just as catastrophic in vision, wrote in 1879 that the moon had been born by being thrown from the fast-spinning earth. The void left by the moon's departure had to be filled somehow, and the Reverend Osmond Fisher in 1882 subsequently suggested that lateral movement of the continents partially accomplished the task. The previous year Fisher, in his book *Physics of the Earth's Crust,* had also offered a remarkably modern description of the earth's structure, with a fluid layer sandwiched between a core solidified by great pressure (or some other cause) and crust made solid by cooling. Fisher also recognized two types of crust, which would prove important in the later debate over continental drift. One type, made of the relatively less dense granite, forms the continents while the other type, composed of basalts, forms the floors of the oceans. The difference in elevation between continents and ocean basins could thus be explained by materials of different density floating atop Fisher's liquid layer. Fisher had thus laid the basis for the concept of isostasy, which proved influential to Wegener's gestating ideas thirty years later.

Heinrich Wettstein, a Swiss schoolteacher, proposed a drift theory in which continental movements were driven not by some fantastic catastrophe but by the gravitational effects of the sun. Unfortunately, his hypothesis was included in a book that contained many "inanities," as Wegener later wrote, and Wettstein regarded the ocean basins as sunken continents. Thus his would-be contribution stumbled into scientific and historical obscurity.

An American geologist, Frank Bursley Taylor, hinted at a theory of continental drift in 1898 in a privately published booklet titled *An Endogenous Planetary System.* Taylor suggested that the moon had been a comet captured by the earth. As a result of the capture, the speed of the earth's rotation would have increased, and the greater centrifugal forces would have pulled continental landmasses closer to the equator. Mountain belts would have arisen on the sides of the continents nearest the equator as the drift occurred. The most complete and widely available presentation of Taylor's ideas came at a 1908 meeting of the Geological Society of America in Baltimore. The print version appeared in the *Bulletin of the Geological Society of America* in 1910. In that paper he attributed the rise of the massive mountain belt that parallels the equator—including the Alps, Atlas, Caucasus, and Himalayan chains—to the drift of continents away from the poles. He asserted that the process was driven by centrifugal forces but made no mention of the moon's capture at that time. He returned in 1923 to an idea of an astronomical origin—the moon's capture—and discussed it in several publications afterward.

Taylor's ideas were similar in some respects to what Wegener would propose in 1912, and Taylor published first, but his hypothesis was not as comprehensive, nor did he summon much evidence outside the realm of traditional geology. An accepted member of the geological community with a number of well-regarded, traditional geological treatises to his credit, he could afford to come up with a strange idea every now and then. Thus, his writings on drift generated little controversy in North America and even less attention elsewhere. Taylor, twenty years older than Wegener, struggled in the years after Wegener's death to maintain a claim on a concept that briefly—in the United States at least—shared both men's names. Even though he merited an obituary in the esteemed American journal *Science* after he died in 1938, Frank Bursley Taylor was largely forgotten.

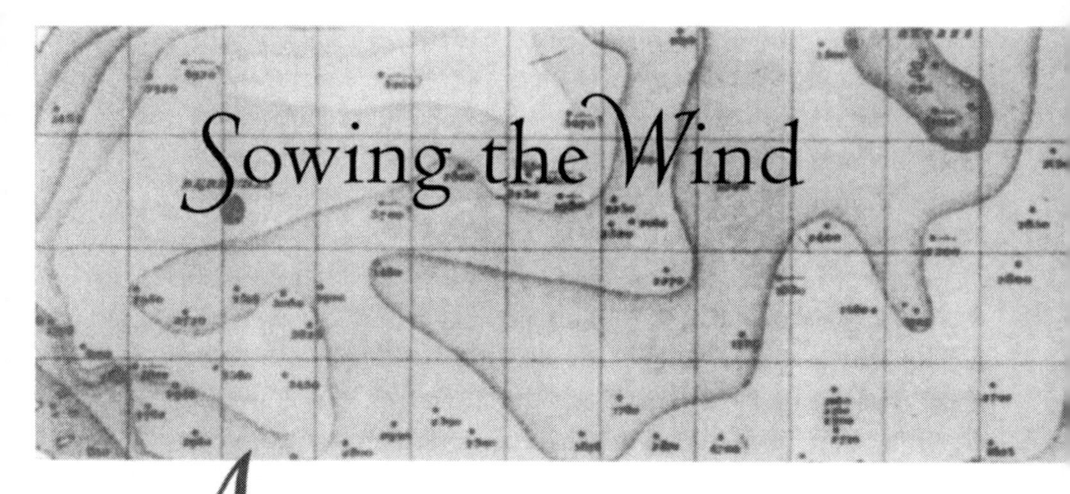

Sowing the Wind

Alfred Wegener descended from a long line of Protestant ministers who worked in Brandenburg and Silesia in what is now eastern Germany and southwestern Poland. Wegener's father, Richard, served as director of the Schindler Orphanage in Berlin. He also taught at the Gymnasium zum Grauen Kloster in Berlin, a school whose most famous graduate was Otto von Bismarck, prime minister of Prussia from 1862 to 1890 and founder and first chancellor of the German Empire from 1871 to 1890.

In 1886 Richard Wegener and his wife, Anna Schwarz, bought a small vacation home in Zechlinerhütte, a small town among the hills and lakes north of Berlin. The young Alfred, along with his older brother, Kurt, took advantage of the idyllic setting to forge a love of adventure and the outdoors. The pair spent as much time as possible hiking and skiing, and as they grew older the brothers sought new challenges.

One of the hobbies the brothers took up was ballooning. Both Kurt, who was also a meteorologist, and Alfred were interested in the research uses of kites and balloons, as well as in the experience of flight itself. The pair's exploits made international headlines when, from April 5 to 7, 1906, Kurt and Alfred set a world record for time aloft in a balloon on a flight across Germany and Denmark. The brothers were poorly equipped for such a long flight, which shattered the previous record by seventeen hours. Only their aeronautical skill and excellent physical condition enabled them to stay aloft so long.

Alfred Wegener's experience with kites and balloons for meteorological research played a role in his getting named to the Danmark Expedition to Greenland in 1906.

AFTER HIS SECOND trip to Greenland, on the Koch expedition, Wegener served in World War I as a reserve lieutenant with Germany's Queen Elizabeth Grenadier Guards. He was not the type to avoid hazardous duty, however, and was shot twice in 1914 while fighting in Belgium. The second wound—in the neck—put him out of combat for the duration of the war. After a lengthy sick leave, Wegener spent the remaining years of the war in the military weather service. Afterward he resumed his scientific career.

Trained in astronomy and physics, but in practice more a meteorologist, Wegener worked in diverse disciplines. In 1911 he wrote a classic textbook on the thermodynamics of the atmosphere—a text that, considering the topic, was notable for the paucity of complex mathematical equations. In 1921 he entered an important scientific debate by publishing a prescient monograph, *The Origin of Moon Craters.* Wegener showed, through a series of experiments with powdered cement, that the craters were most likely created by meteorite impacts rather than bursting bubbles of magma, tidal action, or volcanic action—explanations that were widely discussed at the time. Wegener and his father-in-law, climatologist Wladimir Köppen, weighed in on paleoclimatology in their 1924 book, *Climates of the Geological Past.* Their analysis offered crucial support to Milutin Milankovitch's theory that variations in the earth's orbit, which affected the amount of solar radiation reaching the atmosphere, played a role in the rise and fall of the ice ages. In the years before his second expedition to Greenland, Wegener developed another theory, which wasn't received very well by the scientific community. That theory was called continental drift.

WEGENER'S involvement with continental drift began with a map. In 1910, while a lecturer in astronomy and meteorology at the University of Marburg in Germany, he observed a remarkable similarity in the shape of coastlines on either side of the Atlantic Ocean. The observation stirred his imagination. "Please, look at a map of the world!" he wrote to his then-fiancée, Else Köppen, late in 1910. "Does not the east coast of South America fit exactly with the west coast of Africa as if they had formerly been joined? The correspondence is still better if one compares not the present coasts but the lines of descent to the deep sea." At first Wegener did little to follow up on his observation. But he had diverse interests and found it difficult to confine himself to the library sections on astronomy and meteorology, the fields he lectured in. In the fall of 1911 he found a paper on the similarity of fossil plant and animal species from either side of the Atlantic. That seemed interesting, so he responded as a curious man—but not as would a "respectable" scientist who knew his field and stayed in it. He began looking for other instances of similar biota separated by vast oceans. How did that happen? Was it possible that the continents were joined in the past?

Wegener knew such ideas would be met with considerable skepticism. Instead of keeping quiet, however, he gathered all the evidence he could marshal in support of his theory. Quite a lot was already on hand: similar animal and plant species, fossils, and geological strata found in regions separated by wide, deep ocean basins, such as marsupials in both South America and Australia, suggesting that the two continents had once been joined and drifted apart, and extensions of the Appalachian Mountains in Scotland, suggesting that North America and Europe likewise had been joined and have since separated. On the other hand, fossils from places in which no similar organism can exist today, such as fossils of tropical plants found on Antarctica and on islands in the Arctic Ocean, suggested those lands had drifted from the warm tropics toward the frigid poles.

Wegener first presented his ideas on continental drift on January 6, 1912, in an address, "Die Herausbildung der Grossformen der Erdrinde (Kontinente und Ozeane) auf geophysikalischer Grundlage" (The geophysical basis of the evolution of the large-scale features of the earth's crust [continents and oceans]), before the Geological Association in Frankfurt am Main. His ideas were not well received. Four days later he delivered a similar talk, "Horizontalverschiebungen der Kontinente" (Horizontal displacements of the continents), before a more sympathetic audience at a meeting of the Society for the Advancement of Natural Science in Marburg. Two papers followed later that year, in serialized form in *Petermanns Mitteilungen* and in a single installment in *Geologische Rundschau.* Further development of his ideas was interrupted by the Koch expedition to Greenland and by the outbreak of World War I. Wegener's combat career may have been cut short by the gunshot wound in the neck, but the convalescence afterward gave him time to ponder the motions of the continents. In 1915 he published a book, *Die Enstehung der Kontinente und Ozeane* (The origin of continents and oceans), that presented his theory in detail.

Wegener's argument was simple: About 300 million years ago, all large landmasses were united as one supercontinent, Pangaea. Beginning at about 150 million years ago, Pangaea began breaking up. The fragments drifted apart and, in some cases, collided again, eventually becoming the continents we recognize today. (Wegener also postulated that the process of division and collision had been active before Pangaea came into existence but found it difficult to reconstruct pre-Pangaea geography.) Wegener produced maps of how he thought the continents had been assembled or disassembled at various stages in time. He presented the geological, geographical, climatological, and paleontological evidence in support of his theory, and he explained why competing theories did not work.

Contraction theory was Wegener's first victim, and he scuttled it by reviewing geophysical studies of the nature of oceanic and continental crust. According to most versions of the contraction theory, continental crust and oceanic crust were virtually identical in composition. The only major difference was in their elevation. If one could prove that continents and ocean basins were made of different materials, then one could dispense with contraction. Wegener began his assault on geological orthodoxy with a simple comparison: The average elevation of continents was about 2,300 feet above mean sea level, while the average depth of the ocean floor was about 14,100 feet below mean sea level. In most cases the outer portions of continental platforms slope gently downward, with the lowest reaches covered by ocean waters. At the edge of most continental "shelves," however, a steep escarpment signals an immense drop into the ocean basins. If the drying apple concept held true, the slopes from continental heights to oceanic depths should be more gradual. Defenders of contraction argued that subsidence of ocean basins occurred along extensive fault lines at the edges of the continental platforms, and that these faults explained the steep gradients. If the crust of the continents and oceans were made of similar material, the argument might be valid, but Wegener had plenty of reason to believe otherwise.

First, Wegener cited gravity studies supporting Osmond Fisher's assertion that the two types of crust were fundamentally different. Gravity, in general terms, is the attraction of two bodies toward one another. The strength of attraction is a function of the mass of the two bodies and the distance between them. If the two bodies differ in mass, the smaller body will be drawn toward the larger one; the speed at which it moves toward the larger body will increase (accelerate) as the distance between them decreases. In physics classes the acceleration due to gravity is often treated as a constant. On the surface of the earth, however, gravitational acceleration varies as a result of the earth's shape and rotation. The earth's shape is not a perfect sphere but is slightly squashed:

its diameter is greater at the equator than between the poles (in other words, sea level is "higher" at the equator). The rotation speed of the earth's surface is greatest at the equator and decreases toward the poles (in effect, the centrifugal force trying to throw objects "off" the surface of the earth is greatest at the equator). As a result of the combined effects of the earth's shape and surface rotation speed, gravitational acceleration at sea level will tend to increase from the equator to the poles. Local factors also influence gravity. Gravitational acceleration will generally decrease at higher elevations and increase at lower elevations. If elevation is held constant, the acceleration will be stronger in the vicinity of very dense rock than it will be adjacent to loosely consolidated sediments or water.

In 1666 Robert Hooke suggested that a pendulum could be used to measure gravity. In theory, if one knows the length of the pendulum and the period of its swing (the time it takes for the pendulum to swing from the right to the left and back again to the right), then one can calculate the acceleration due to gravity. In practice, however, measuring the period of a pendulum with sufficient precision to reveal regional variations in the force of gravity is not easy. For example, a one-thousandth of a second error in the measurement of the period of an eighteen-inch-long pendulum could mean the difference between the average strength of gravity at the equator versus the strength of gravity at about thirty-two degrees latitude north or south. Obtaining precise measurements on land was difficult enough, as mistakes in timekeeping and in counting the swings of the pendulum were fairly easy to make. But measuring gravity at sea was next to impossible because ship motions also affected the motion of the pendulum.

Around the turn of the twentieth century, scientists developed another way to measure gravity at sea using a type of hypsometer that combined a mercury barometer and a boiling-point thermometer. The height of a column of mercury in a barometer is affected by two things: atmospheric pressure and gravity. Water boils at lower temperatures as atmospheric pressure decreases.

Thus, by precise measurement of the boiling point of water, a scientist on a ship could determine atmospheric pressure at sea level. By isolating the effect of pressure, the scientist could then use the height of the mercury column to calculate the local acceleration due to gravity.

The rocks of the earth's crust are more than twice as dense as water, so nineteenth-century scientists expected to find that gravity would be weaker over the oceans than it was over the continents. By the time Wegener began developing his ideas on continental drift, however, it was known that the supposed gravity deficit did not exist. The obvious conclusion was that the crust of the ocean floor was denser than the crust of the continents. That in turn meant that oceanic rocks must be made of different materials than continental ones. The dried-apple concept of contraction began to shrivel as such evidence proved that oceanic and continental crusts were different.

Another implication of the gravity findings made the dried-apple crumble away entirely. The gravity results lent credence to the then-controversial theory of isostasy—which, unlike the dried-apple theory, suggested that oceanic and continental crusts must be different.

Wegener was ahead of many traditional geologists of the day in his comprehension of isostasy—a theory developed in large part by Fisher and named by American geologist Clarence Dutton, who along with another American, Grove Karl Gilbert, also contributed greatly to its development. Isostasy builds on Archimedes's principle which states that a body floating in a fluid is supported by a force equal to the weight of the displaced fluid. The principle explains why a block of more dense material floats lower in a buoyant medium than a block of a less dense substance—as, for example, a freighter will ride lower in water when fully laden than when empty because a greater force is required to keep the fully loaded ship afloat. A greater force means more water is displaced by the full ship, which drops deeper (rides lower) in the water in order to fill the void created by the displaced water.

Isostasy stipulates that structures on the earth's surface will tend to rest at an elevation where the buoyant and displacement forces are balanced. As mountains are eroded at the top, the underlying structure will rise like a freighter being unloaded, such that elevation of the peaks will remain virtually unchanged. Continental areas such as Hudson Bay, once pressed down by the weight of massive ice sheets as much as two miles thick, will rebound in height as the glaciers melt. Depressions filled by a constant rain of sediments will sink further under the weight of those sediments without noticeable change in the elevation of the surface. Wegener noted that evidence for isostasy was widespread by the time he conceived of continental drift. To him, it was clear that the average elevation of continental platforms was higher than that of the ocean floors because the material that made up the continents was less dense than the material that made the ocean floors.

By the early part of the twentieth century, geologists had determined that the most common elements in continental rocks were silicon and aluminum. Many volcanic rocks were made primarily of silicon and magnesium. Early attempts to sample the composition of the ocean floor with dredges also brought up rocks high in silicon and magnesium. Eduard Suess coined the terms *sima* to describe volcanic rock and *sal*—later referred to as *sial*—to describe the more common material that made up the continents. (Suess also coined the term *nife* to describe the nickel-iron core of the earth, *ferrum* [fe] being the Latin word for iron.) The specific gravity (ratio of the density of a substance to the density of water) of sial ranged from 2.5 to 2.7; the specific gravity of sima ranged from 3.0 to 3.3. Again, from the sampling evidence, oceanic and continental crust seemed to be composed of different materials.

Wegener had more discouraging words for supporters of contraction theory. If the dried-apple concept held true, then the wrinkles should be more evenly distributed across the surface of the earth rather than concentrated, as they are, in massive belts of folded and overthrust mountains. Wegener also thought it improbable that contraction could produce wrinkles in some cases but sequences of massive uplift and subsidence in others. In a further blow to the theory of uplift and subsidence, he noted that sedimentary strata on continental platforms typically contain materials laid down in shallow seas rather than material of the abyssal depths.

The well-read Wegener noted the significance of radioactivity in the earth. While the newly discovered source of energy dealt a blow to Thomson's 20-million-year-old earth, it also cast doubt on the premise that the planet had cooled enough to create all the features contraction theory was said to explain. Wegener also correctly interpreted rift valleys as areas where the earth's crust was splitting apart—not what one would expect to see on a shrinking planet. The studies of gravity, isostasy, and the specific gravity of the materials that made up the outer layers of the earth, coupled with the additional evidence, clearly indicated that continental and oceanic crust were different. One conclusion was obvious: Contraction theory was dead.

Wegener then turned to geology to prove that the continents had, in fact, moved across the face of the earth. The clearest evidence for his argument came from the Southern Hemisphere, where Wegener pointed out how well Africa and South America would fit together along their Atlantic shores. Wegener looked deeper than the coastlines, however, noting that the shape of a coast may change as sea levels advance and recede. Thus the shoreline at any one time should not be considered the edge of a continental platform. Instead, he correctly deduced that the edges of the continental shelves mark the boundaries. When Wegener compared the shapes of continental shelves, the fit between the African and South American landmasses—and others—improved markedly.

While Wegener was impressed by the fit, it wasn't enough by itself to convince him (or many other scientists) that the continents had drifted. However, if he could line up other features of the continents by joining them together, he thought he would be able to prove his case. "It is just as if we were to refit the torn pieces of a newspaper by matching their edges and then check whether the lines of print run smoothly across," Wegener wrote. "If they do, there is nothing left but to conclude that the pieces were in fact joined in this way." By searching the geological literature, he found plenty of evidence that a connection had existed. For instance, the Cape Ranges of South Africa have their counterpart in the Sierras in southern Buenos Aires Province in Argentina. The gold-bearing granites of the Minas series in the Brazilian states of Minas Gerais, Goiás, and São Paolo are similar to those of the Bushveld complex in the Transvaal of South Africa as well as the Erongo granite of Heroroland and the Brandberg granite of Damaraland in what is now Namibia. Supporting evidence came from other regions as well. The northern Appalachian–East Greenland mountain system has its counterparts in the Caledonian mountain belt of the British Isles and Scandinavia. The mountains formed during the Alleghenian orogeny—or Alleghenian mountain-building event—in what is now eastern North America are matched by the mountains formed during the Hercynian orogeny in northwestern Africa and western and central Europe.

Geological matchmaking wasn't the only method used by Wegener to prove continental drift. The island of New Guinea—which Wegener correctly interpreted as geologically part of Australia—is flanked by two curiously curved island arcs, the Bismarck Archipelago to the east and the Lesser Sunda (Nusa Tenggara in Indonesia) and Molucca Islands to the west. Wegener suggested that the island chains were bent as the prow of New Guinea thrust between them. Another curious arc cited by Wegener lies between Graham Land on the Antarctic Peninsula and Tierra del Fuego at the tip of South America. Both Graham Land and Tierra del Fuego gently curve eastward. The South Sandwich Islands lie at the apex of the arc, with South Georgia along the northern rim and the South Shetlands, Elephant Island, and the South Orkneys along the southern rim. Wegener suggested that, as Antarctica and South America drifted west, the various island chains became stuck in the sima that made up the ocean floor.

The next battery of evidence Wegener drew upon came from biology and paleontology. In addition to the fossil *Mesosaurus* and *Glossopteris* evidence, he saw support for continental drift in the distributions and habits of living organisms. For example, one family of earthworms, the Lumbricidae, occur in a belt across Eurasia from Japan to Spain as well as in eastern North America (but not western North America!). Lumbricids cannot survive in salt water, nor can they cross frozen ground. He believed the distribution could be explained only if North America and Europe had once been joined. Eels, too, were featured in Wegener's defense of drift. Eels spend much of their lives in the rivers of Europe and North America, but they breed in the Sargasso Sea, in the western part of the North Atlantic Ocean. This again struck Wegener as solid evidence that the two continents had once been joined. While many other scientists continued to argue for the existence of sunken continents or land bridges, Wegener in most cases dismissed the allegedly sunken landmasses; however, he did believe in the Bering Strait land bridge between Alaska and Russia. The principle of isostasy would not allow continents or most other land bridges to disappear so conveniently below the waves. The only logical explanation, then, was that the continents, at one time joined, had drifted apart.

Rocks and fossils provided a key to understanding the environments of the past, and Wegener was a pioneer—independent of his work on continental drift—in uncovering the secrets locked in a sequence of stone. Smooth, rounded stones are created by the gentle force of water; thick beds of salts, such as gypsum, are usually formed by hot, dry climates where the little water available evaporates to leave the crystals behind; limestones indicate warm, shallow seas; coal and oil indicate a warm, moist environment favoring lush vegetative growth; thick piles of debris where particles of all sizes, from silt to huge boulders, are mixed often indicate deposition by a glacier. Fossils were also useful. Trilobites occurred in marine environments; fish with fins modified for crawling indicated arid areas where ethereal pools may dry up; bees in amber suggested the presence of flowering plants.

Wegener bolstered his argument for continental drift with his experience studying the climates of the past. One of the first lines of paleoclimatic evidence he cited concerned an ice age during the Permian and Carboniferous periods from 225 million to 385 million years ago. Glaciers, in addition to leaving behind piles of unsorted debris as they melt and retreat, create characteristic grooves in rock they grate against. Virtually no signs of the glaciation are found in the Northern Hemisphere, but ample evidence is found south of the equator. Geological evidence for the Permian-Carboniferous ice age crops out in unlikely places, from Australia, Africa (as far north as the Congo), South America (as far north as Brazil), and peninsular India. Wegener's analysis indicated that, without continental drift, the South Pole would have had to be in the Indian Ocean southeast of the Cape of Good Hope. Wegener concluded that drift was much more likely. Former ice ages weren't the only phenomena invoked to support his theory. Coal seams in Antarctica (from which the fossils of the *Glossopteris* flora were recovered) indicated a warm, moist climate. Without continental drift, temperatures would have had to be searing—unlivable for almost any organism—in the tropics for the now frozen Antarctica to support a lush, swampy forest. But the fossils could easily be explained by a continent drifting away from the equator and toward the South Pole.

In his effort to build an all-inclusive theory, Wegener was bound to stumble. He diluted his argument for continental drift by ascribing some shifts in climate zones to polar "wandering"—or shifts of the entire surface of the globe with respect to the spin axis. He also placed too much faith in estimates of continental movements based on traditional methods of celestial navigators. By shooting the elevation angle of known objects in the sky, obtaining the precise time, and making the appropriate calculations, a skilled observer can figure out where he or she is located on the earth with remarkable accuracy. Danish expeditions had repeatedly tried to measure the longitude of Greenland by observing the position of the moon. The technique—at one time seriously considered by Britain's Royal Navy in a desperate search for a way to accurately determine longitude, but which fell victim to John Harrison's magnificent chronometer—was fraught with error. Measuring elevation angles of stars with sufficient precision is relatively easy—all the observer has to aim at is a small dot in the sky. The moon, however, is a huge target. It may be easy to spot, but for navigation or geodetic purposes the observer has to measure the angle of the same feature on the moon—such as the bottom of the lunar disk—time after time. Such repeated acts of pinpoint accuracy are virtually impossible to achieve. Time itself is a problem. Stars, because of their distance from the earth, seem more or less motionless in the sky, so that slight errors in measuring the time don't normally amount to large errors in location. The moon, in contrast, makes a relatively speedy transit across the heavens. Accurate time is essential to accurate location finding based on the moon. Finally, longitude measurements made over a period of years have to be made at the exact spot to be of value. Years, sometimes decades, separated expeditions to locations in Greenland from which the key celestial observations were made. Finding the right exact spot years after a previous visit, especially in the island's harsh climate, is not easy. Despite their best efforts to control for these sources of error, Danish observers had estimated that Greenland drifted to the west (away from Europe) at a frighteningly brisk rate of thirty-six meters per year. Wegener enthusiastically embraced the findings as proof of drift. Unfortunately, in the years following Wegener's death, the Danes discovered their error. The technique failed to show that Greenland had moved at all.

Wegener was met with howls of derision in his conception of the continents as analogous to icebergs: the rigid but less dense blocks of sial floated in a "sea" of sima (the denser material of the ocean floor), and as the sial blocks drifted along they pushed the sima aside. The comparison made sense to the polar explorer familiar with real icebergs, but most other scientists could not envision how the hard rocks of the ocean floor could give way so readily. Again, Wegener drew upon his polar experiences. Glaciers are made of massive sheets of hard, solid ice, yet under great pressures they can act as a fluid, flowing around immovable obstacles such as isolated mountain pinnacles. Wegener correctly pointed out that solid rock can behave as a fluid under sufficient pressure. While the pressures at the ocean floor are indeed tremendous, however, they aren't enough to produce fluid motion in the rocks.

Probably the biggest problem with Wegener's theory was the question of what mechanism drove the motion of the continents. He considered several mechanisms, such as a general westward drift caused by tidal friction and a force driving landmasses away from the poles toward the equator (similar to what Frank Taylor had proposed). Wegener toyed with Darwin's and Fisher's idea of the birth of the moon from the Pacific Ocean basin for a time but had dropped it from consideration by his fourth and final edition of *The Origin of Continents and Oceans.* Wegener even considered convection currents deep inside the earth. Not enough was known about convection, however, and the westward drift and pole-flight force, even in combination, were too weak to account for everything Wegener sought to explain. He readily admitted that the uncertainty about the mechanism of drift was a problem. "The Newton of drift theory has not yet appeared," he wrote in the fourth edition of *The Origin.*

Wegener had no doubt about the importance of drift theory, however. Again in the final edition of *The Origin of Continents and Oceans,* he wrote, "We may, however, assume one thing as certain: *The forces which displace continents are the same as those which produce great fold-mountain ranges.* Continental drift, faults and compressions, earthquakes, volcanicity, . . . and polar wandering are undoubtedly connected causally on a grand scale."

Reaping the Whirlwind

Wegener's timing left something to be desired. Early in 1912 he introduced a radical new theoretical framework for explaining the earth—one that invoked tremendous horizontal displacements of the continents yet did not require (but could accommodate) planet-shattering catastrophes that were occasionally discussed in the scientific community at the time—then departed that summer for the expedition to Greenland with Johan Peter Koch. About a year after his return to Germany following the expedition's harrowing, but successful, crossing of the island in 1913, Wegener was called to active duty in the German army as World War I broke out. Naturally, he had little to say on the subject of continental drift as he led troops into battle on the western front, but other German scientists managed to take note of his ideas as the military conflict erupted across Europe and the Middle East.

The initial reaction among German scientists was mixed. Karl Andrée, in a monograph published before the war in 1914, wrote that he agreed with some aspects of Wegener's theory but still had difficulty accepting large-scale movements of continental blocks, nor could he envision a force capable of driving such movements. Carl Diener in 1915 adamantly endorsed the permanence concept and added that Wegener's theory was nothing more than an attractive mind game. Wilhelm Soergel, deeply disturbed by how easily Wegener's proposal could undercut the theoretical basis of paleogeography at the time, presented in 1916 a number of arguments why the Atlantic Ocean could not have been formed by the separation of the Americas from Europe and Africa. He urged rejection of Wegener's ideas.

Max Semper, however, eclipsed the other German critics with a vehement attack on Wegener and his theory. "It is certain that Wegener's theory was established with a superficial use of scientific methods, ignoring the various fields of geology," Semper wrote in 1917. "It is easy to prove that each occasion for high-flying nonsense was fully and successfully used, that the arguments rest either on misunderstanding or cannot prove what they are supposed to, and that, finally, what should have been considered, was almost regularly ignored."

Semper did not stop at slamming Wegener. He had something to say about those who found some favor with continental drift as well. "It would not have been necessary to go into so many details about this unfortunate attempt to 'pull a trick on the Earth,'" Semper wrote, "had it not in the meantime acquired supporters who either did not notice how poorly documented it was or who considered it worth discussing anyway."

Among those who had quickly embraced Wegener's "trickery" were Edgar Dacqué, who by 1915 had recognized the almost all-encompassing explanatory power of the theory of continental drift; Andrée, who had published a paper more supportive of Wegener's ideas in 1917; Emile Argand in Switzerland; and Gustaaf Adolf Frederik Molengraaff and Hendrik Albertus Brouwer in the Netherlands. Argand, Molengraaff, and Brouwer all had experience in critical regions: Argand in Switzerland, where the Alps—according to Wegener's theory—had been formed by a collision between Europe and Africa; and Molengraaff and Brouwer in the Dutch East Indies and Southeast Asia, where continental collisions could account for the volcanic islands and bizarrely shaped Indonesian and Bismarck Archipelagoes. Two other factors were also key to their early awareness and support of Wegener's ideas. First, they were not greatly hindered by language differences. Second, they were from countries that remained neutral during the war. (Argand's advocacy proved risky, for when he gave a talk on Wegener's theory before the Neuchâtel Society of Natural History, anti-German feeling was rampant in Switzerland, and—even more important—reading material printed in Germany was prohibited!)

The four years following the end of World War I were a critical time in the development of the theory of continental drift. Wegener published a second edition of *The Origin of Continents and Oceans* in 1920. He had largely silenced his German critics during a debate in Berlin in 1921 and then published a third edition of *The Origin* in 1922. The third edition was translated into English, French, Russian, Spanish, and Swedish. As a result, more and more scientists became aware of Wegener's ideas—not always to his benefit.

WEGENER'S theory did not fare well in France and Belgium, the two countries that bore the brunt of the fighting on the western front during the Great War. Both had been exhausted by the war, and many young members of their respective scientific communities had perished in the fighting. Elie Gagnebin, a Swiss geologist born in France, introduced the French to Wegener's theory with a paper in the journal *Revue Générale des Sciences Pures et Appliquées*, Paris, in 1922. Gagnebin found it rather ironic that Wegener's theory was largely ignored in a nation where the ideas of Einstein and Freud were widely discussed. Gagnebin was troubled by Wegener's style of presentation but found that the fundamental ideas could not be ignored. Despite Gagnebin's introduction, however, mainstream French scientists, with few exceptions, had a difficult time comprehending the theory. For decades little of the French debate concerning drift descended from the lofty heights of grand ideas—where participants generally remained ambivalent—to the lowly empirical world where the theory of continental drift was put to the test. The few French scientists who managed to fall to earth, such as John Leuba and Philibert Russo, found much to recommend in Wegener's ideas.

The Belgians, unlike their French counterparts, wasted no time on ambivalence. Under the guidance of François Kaisin of Louvain and Paul Fourmarier of Liége, they rejected continental drift outright.

The Austrian reaction to Wegener was more complex and dynamic. Vienna was the home of Eduard Suess, who had won great renown for his contraction theory. However, two other Austrians, Otto Ampferer and Robert Schwinner, independently developed theories on mountain building and volcanism that depended at some level on horizontal displacement of the continents. Regardless, the influential Viennese geologist Leopold Kober was a staunch opponent of drift. Most of his Austrian colleagues followed his lead but, under the tutelage of their enthusiastic Swiss neighbors, slowly yielded to drift by 1927.

The lack of an Italian edition of *The Origin of Continents and Oceans* hindered acceptance of Wegener's ideas on the peninsula. While some reviews were favorable, Federico Sacco—who, ironically, had in 1906 developed a drift hypothesis of his own based on contraction theory—spoke for the Italian establishment in a 1919 paper "Aberrazioni" (Aberrations). The title adequately summed up his opinion of Wegener's theory.

The Spanish, who also remained neutral during World War I, were, like the Dutch and Swiss, open-minded and enthusiastic about Wegener's ideas. Lucas Fernandez Navarro, in a 1922 review of the second edition of *The Origin of Continents and Oceans*, cited some problems with the theory of continental drift but was impressed by its explanatory power. Navarro wrote, "One cannot deny the fact that it simplifies the understanding of many phenomena which until today have remained unexplained or were inadequately accounted for." In a 1925 speech at the Royal Academy of Science in Madrid, Navarro said, "The theory by Wegener is the dawn of new promises in geology. We should view these new ideas with sympathy as we look at young people full of promises and hope."

The residents of Albion were as perfidious as usual in their response to continental drift. The first reference to Wegener's theory in British literature appeared in November 1913 in a review section of *Geographical Journal.* Despite the early introduction, drift received little official notice until 1922, when the second edition of *The Origin of Continents and Oceans* was reviewed in the February 16 issue of *Nature.* The anonymous reviewer noted that the theory was generating a significant amount of opposition among geologists, for it "replaces the whole theory of sunken continents, land bridges, and great changes of earth temperature by a displacement theory." The reviewer, however, realized the potential of Wegener's theory to overturn current theories about the earth. "The revolution in thought, if the theory is substantiated," the reviewer wrote, "may be expected to resemble the change in astronomical ideas at the time of Copernicus."

The first shot of the revolution in Great Britain, such as it was, began in Manchester with an article in the March 16, 1922, edition of the newspaper *The Guardian,* "The Displacement of Continents: A New Theory." The article, by F. E. Weiss, a professor of botany at Manchester University, reviewed Wegener's theory and found it promising. Five days later Weiss and geologist W. B. Wright discussed continental drift at a meeting of the Manchester Literary and Philosophical Society. In May Wegener himself wrote an article on continental drift in the English magazine *Discovery.*

The establishment counterattack began with a special meeting of the Manchester Literary and Philosophical Society on May 2. Sir William Boyd Dawkins, a former president of the society from 1903 to 1905, reviewed the state of geological theory at the time and found to his satisfaction that the foundation of the science was unshaken by Wegener's challenge.

While Wegener's theory was gaining support on the Continent following Gagnebin's review and a tour-de-force presentation by

Argand on the tectonics of Asia at the International Geological Congress in Brussels, Belgium, in August 1922, the first full-scale assault on Wegener began in England in the August 1922 issue of *Geological Magazine.* Philip Lake, of the Sedgwick Museum in Cambridge, highlighted mistakes, or perceived mistakes, in Wegener's theory. He questioned Wegener's honesty. "Wegener himself does not assist his reader to form an impartial judgment," Lake wrote. "Whatever his own attitude may have been originally, in his book he is not seeking truth; he is advocating a cause, and is blind to every fact and argument that tells against it." While Lake had no difficulty accepting vertical movement of continental blocks by isostasy, he did not see how they could move horizontally except through some catastrophe such as the moon being cast off from the earth.

Wegener's theory took another beating in the British Isles in 1923. Lake spoke before the Royal Geographical Society on January 22 in his most vehement attack yet. The proceedings were published in the March issue of *Geographical Journal* and included comments by seven other scientists who had spoken up during the meeting. Six had something positive to say about continental drift. The seventh, however, was damning. Geophysicist Harold Jeffreys wasted no time in exploiting what Wegener had long acknowledged as probably the weakest part of his theory: mechanism. Jeffreys, like William Thomson before him, had harnessed the power of mathematics and found the two primary forces Wegener invoked to explain the continental motion—the westward drift and the flight from the poles—wanting. Jeffreys elaborated on his criticism of Wegener in a letter to *Nature* on April 23. Jeffreys also doused Osmond Fisher's idea that the birth of the moon pulled the Americas toward the west, figuring that if the moon had been created by fission from the earth, it would have had to happen too early in the earth's history to have any effect on the distributions of fossil plants and animals.

In 1924 Jeffreys published *The Earth: Its Origin, History, and Physical Constitution,* a book in which he labored, successfully, to put geophysics on a sound mathematical footing. He also used the opportunity to blast away at Wegener. In addition to his calculations that showed Wegener's mechanism was inadequate, he introduced evidence that the crust of the earth was too strong to permit displacement and deformation on a scale envisioned by Wegener. However, he took his logic one step too far. Jeffreys may have proved that Wegener's proposed mechanism was insufficient—and he stood on solid ground as far as that went—but he went on to say that one could not accept drift until a plausible mechanism was found. Wegener had realized that proving *whether* drift had occurred was one problem, while proving *how* it occurred was another. Jeffreys hopelessly confused the two. Unfortunately, he managed to infect the majority of scientists in Britain and the United States with his addled thinking on that point.

JOHANNES GEORGI, the former student of Wegener's who accompanied him to Greenland in 1929 and 1930, discussed what was probably the most serious problem Wegener faced throughout his career. In a contribution to a 1962 volume on continental drift, Georgi wrote:

> It may be mentioned in these personal reminiscences how very much Wegener's colleagues regretted the fact that this great scholar, predestined for research and teaching, could not get a regular professorship at one of the many universities and technical high schools in Germany. One heard time and again that he had been turned down for a certain chair because he was interested also, and perhaps to a greater degree, in matters that lay outside its terms of reference—as if such a man would not have been worthy of any chair in the wide realm of world science.

Supporters and critics of Wegener had a difficult time deciding what he was. Wegener's doctoral degree was in astronomy; however, he published research in geography, geology, geophysics, and meteorology. The effects of this intellectual ambiguity were somewhat humorous: Whatever a commentator might call himself, he made sure to call Wegener something else. A physicist might call Wegener a geologist. A geologist might refer to Wegener as a meteorologist. A meteorologist—even Wegener's father-in-law, Wladimir Köppen—might label him a geophysicist. It did not seem to matter what Wegener was, as long as he was not one of "us."

HOSTILE HARDLY DESCRIBES the reaction to Wegener's theory in North America, especially in the United States. Many felt that he should have stuck to astronomy and meteorology. Some of the attacks stepped outside the boundaries of polite scientific discourse. William Berryman Scott, president of the American Philosophical Society, dismissed the theory as "utter, damned rot!" in 1923.

Willem Antonius Josephus Maria van Waterschoot van der Gracht possessed a lot more than what was arguably the longest name in geology. Van der Gracht, who had traveled and worked all over the world, including in Dutch possessions and North America, had a vast knowledge of the earth. A man familiar with the elegance of Amsterdam and the mystery of the islands of what is now Indonesia, he was, in 1926, living in exotic Ponca City, Oklahoma, while working as vice president of Marland Oil Company (a precursor of today's Conoco). He was also one of the earliest of the few supporters of continental drift theory in the United States.

The American Association of Petroleum Geologists gathered in New York City on November 15, 1926, for their fall meeting. One of the featured events was a symposium on the theory of continental drift. That night, in the auditorium of the Engineering Societies Building, van der Gracht was to defend the concept of continental drift before a less than hospitable audience. Some of the biggest names in American geology were to participate: Bailey Willis; Rollin Thomas Chamberlin of the University of Chicago; Charles Schuchert of Yale University; Chester Ray Longwell, also of Yale; William Bowie of the United States Coast and Geodetic Survey; Charles David White of the National Research Council;

and Joseph Theophilus Singewald, Jr., and Edward Wilber Berry of Johns Hopkins University. Most hoped to crush the Wegenerian heresy. Frank Taylor was present, too, to discuss his version of continental drift. In addition to Wegener and van der Gracht, the Europeans invited either to participate or to submit a paper in the printed proceedings—Molengraaff, of the Institute of Technology in Delft, Netherlands; John Joly of Trinity College in Dublin, Ireland; and John Walter Gregory of Glasgow University in Scotland—were all sympathetic to drift in some form, although not necessarily to Wegener's version.

Van der Gracht opened the discussion with sage counsel:

> The problem of continental drift has raised considerable and spirited discussion in geological circles. Many authorities, entitled to all respect, advocate it; others are undecided but favorably inclined; still others do not favor it, and some are violently opposed. The whole controversy reminds me vividly of the discussions during my student days on the problem of sheet-overthrusting in the Alps. As now in the discussion of continental drift, so there was then much opposition, in which no less an authority than Albert Heim took a leading part before his conversion to the new idea. Its mere possibility was then as firmly denied, as is now the possibility of continental drift. The facts have since proved beyond any doubt that these sheets exist, not only in the Alps, but universally. Still their detailed mechanism, their "possibility," remains almost as much a riddle as it was then. The possibility has only been demonstrated by fact, not explained.

Van der Gracht then went on to warn his audience to not be too satisfied with pet theories. Only an infinitesimally thin slice of the earth's surface could be studied directly; the mysteries of the subterranean world would be revealed only through a collaborative effort of scientists of myriad disciplines. He then showed why one pet theory—contraction—was insufficient to account for all the phenomena ascribed to it.

Next, van der Gracht reviewed the latest research on the structure of the earth. The outer layer, the sial, rested on top of a thicker layer of sima. The density of sial rocks ranged from 2.5 to

2.7, that of sima from 2.9 to 3.0 (although one researcher suggested the density of sima was as high as 3.6). Citing the work of Joly, van der Gracht said that isosasty had been all but proven—thus the less dense sial blocks floated on top of and in hydrostatic equilibrium with the more dense sima. While estimates of the thickness of the outer sial crust varied, most ranged from 20 to 60 kilometers (12 to 37 miles). The crust was underlain by two layers, the inner silicate mantle and the "Pallassite" zone (together corresponding with today's mantle), which extended down to 2,900 kilometers (1,800 miles); and the nucleus (today's core) below 2,900 kilometers.

The composition of ocean floors was a mystery. Van der Gracht reckoned that they were sima in most places (such as under the Pacific and Indian Oceans). However, he felt that an irregular layer of sima covered the ocean floors in the Atlantic and Arctic Oceans.

After describing other characteristics of the earth's layers, van der Gracht took up continental drift. At the time three main theories of drift existed: Taylor's, Wegener's, and one described in 1923 by Reginald Aldworth Daly. Daly had proposed that gravity pulled the continents down bulges in the crust—like "enormous, slow landslides" in Daly's words—created by contraction of the planet. Van der Gracht devoted the rest of his talk to how a suitable mechanism for drift might be found in John Joly's research.

Van der Gracht closed his address with a plea for an impartial exchange of ideas:

> As study proceeds the theory will change, but I am personally convinced that there is continental drift, not necessarily either Taylor's or Wegener's or Daly's, but still inter-continental drift. If the opponents of this view, of whom I see many in my audience, and from whom we will presently hear, will only consent to see it in this light, we may hope to make better progress in gradually approaching the truth.

Willis, in his address, pronounced himself impartial, then proceeded to portray Wegener's ideas as ridiculous. Willis did not waste much time criticizing Wegener's mechanism—where criticism was more than warranted—but instead devoted most of his time to attacking Wegener's evidence. For instance, Willis noted the similarity in shape between the Atlantic coastlines and continental shelves of South America and Africa. To Willis, however, such similarities were not proof that the two continents had once been joined and then drifted apart, which was fair enough, even by Wegener's standards. Willis took his reasoning to new "heights," though, when he said such close correspondence in form were in fact proof that South America and Africa had *not* drifted away from each other. In Willis's mind, the trailing edges of the continents would have been destroyed by massive faulting as the Atlantic opened, thus erasing any resemblance in form!

Willis also took issue with Wegener's presentation and use of "indirect" evidence. "When we consider the manner in which the theory is presented," Willis wrote,

> we find: that the author offers no direct proof of its verity; that the indirect proofs assembled from geology, paleontology, and geophysics prove nothing in regard to drift unless the original postulate of drifting continents be true; that the fields of related sciences have been searched for arguments that would lend color to the adopted theory, whereas facts and principles opposed to it have been ignored. Thus the book leaves the impression that it has been written by an advocate rather than by an impartial investigator.

Again, Wegener couldn't win with Willis. Data on geological strata, fossil flora and fauna, and geophysical characteristics—while considered suitable "direct" evidence in the search for answers to many other geological questions—were nothing more than "indirect" evidence in the debate over continental drift. Furthermore, such data could not be used to prove drift unless drift had already been proved! Finally, in the view of Willis and many others, inappropriate presentation negated the facts of the argument.

Chamberlin came out swinging. "Wegener's theory, which is easily grasped by the layman because of its simple conceptions, has spread in a surprising fashion among certain groups of the geological profession," he wrote. "Other groups of the profession ask: 'Can we call geology a science when there exists such difference of opinion on fundamental matters as to make it possible for such a theory as this to run wild?'"

Chamberlin then pronounced the theory "utterly untenable" and began to present reasons why everyone else should come to the same conclusion. Mercifully, time and space constraints forced him to press his case in outline form only. Some of Chamberlin's criticisms of continental drift were legitimate, but he had clearly not read Wegener's publications closely enough. Chamberlin claimed that continental drift was "no general theory of earth behavior or earth deformation." For Chamberlin it simply described "one supposed breaking up of a consolidated land mass and the migration of the different fragments," despite the fact that Wegener showed how it could explain mountain-building events, volcanism, earthquakes, correspondences of geological strata and fossil beds in now widely separated regions, and the distributions of past climates and of plant and animal groups. Wegener also suggested that drift processes had occurred throughout geologic time but the evidence allowed reasonably accurate description only of events since the existence and breakup of Pangaea.

Chamberlin seemed to fear the revolutionary implications of Wegener's theory. "But taking the situation as it now is," he wrote "we must either modify radically most of the present rules of the geological game or else pass the hypothesis by."

Schuchert employed a globe and plasteline (a type of clay) cutouts in his attempt to do away with Wegener. The cutouts, which matched the shape of the continents at their continental shelves, were shifted and juxtaposed to see how well the results compared with Wegener's reconstruction. Schuchert determined that Wegener's proposal would have left gaps between areas that should have been joined. He also concluded that the continents could not have been joined as Wegener proposed without tremendous distortion; for example, in his reconstruction the Americas would have had to be stretched by 1,500 miles to fit Wegener's scheme. Unfortunately, Schuchert muffed the reconstruction. Had he paid more attention to Wegener's own figures, the fit would have been better. Like Willis, Schuchert doubted that the separating edges of South America and Africa would have retained their shape after hundreds of millions of years of drifting apart. Schuchert also acknowledged the similarities in geologic strata and fossil assemblages on either side of the Atlantic, but—based on calculations he chose not to reveal—concluded that the similarities weren't numerous enough to be evidence for drift.

The most vicious criticism came from Berry. "My principal objection to the Wegener hypothesis," Berry wrote, "rests on the author's method. This, in my opinion, is not scientific, but takes the familiar course of an initial idea, a selective search through the literature for corroborative evidence, ignoring most of the facts that are opposed to the idea, and ending in a state of auto-intoxication in which the subjective idea comes to be considered as an objective fact." So much for just the facts. Berry, ignoring widely accepted concepts such as isostasy and variations of gravity and the different composition of continental and oceanic crust, flatly declared that "the facts of geophysics do not furnish any positive support to Wegener's ideas." Like Willis and Schuchert, he did not think it possible that continents could move thousands of miles apart without destroying the shape of their continental margins. Despite Wegener's work in paleoclimatology, Berry concluded that "Wegener obviously does not know what geological climates were like, nor does he seem to me to be conversant with the established facts of historical geology, since many of his age determinations are erroneous." As for Wegener's paleontological evidence, Berry wrote, "Continental drifting, exactly as in the case of the supposed land-bridges over oceanic areas that it was designed to replace, raises more distributional problems than it solves."

Not all contributors were so negative. Gregory, Joly, Molengraaff, and Singewald professed a belief in continental drift in some form, although, like van der Gracht, they doubted the particulars of Wegener's theory. Longwell was critical of continental drift but called for an open mind among investigators. Taylor and his ideas were virtually ignored by the most vehement critics. His paper added little new to the debate. Neither, for that matter, did Wegener's.

The venomous outbursts of some of his critics at the 1926 symposium illustrated one of Wegener's greatest problems: His unusual, multidisciplinary approach did not sit well with the majority of scientists who, as specialists, were blinded by their expertise in and the limitations of their respective specialties. He did not bring up the issue in his contribution to the symposium but did address the challenge faced by scientists in the preface to the foreword to the fourth edition of *The Origin of Continents and Oceans.*

"We are like a judge confronted by a defendant who declines to answer, and we must determine the truth from circumstantial evidence," Wegener wrote:

> All the proofs we can muster have the deceptive character of this type of evidence. How would we assess a judge who based his decision on part of the available data only?
>
> It is only by combining the information furnished by all the earth sciences that we can hope to determine "truth" here, that is to say, to find the picture that sets out all the known facts in the best arrangement and that therefore has the highest degree of probability.

IN THE YEARS between the 1926 symposium and Wegener's death in 1930, two new geologists began to contribute to the debate, Arthur Holmes and Alexander Logie du Toit. Holmes, a geologist at the University of Edinburgh, was never thrilled with Wegener's methods or style of presentation, nor did he believe in all of Wegener's arguments for continental drift; yet in 1928, in his paper "Radioactivity and Continental Drift," he proposed that convection currents fueled by radioactive heating in the mantle may be responsible for driving drift. On the other hand, du Toit, a South African regarded by some as the "world's greatest field geologist," became continental drift's most ardent defender—although he also recommended modifications to Wegener's original ideas. Unlike most of the critics of drift, du Toit had traveled extensively in the Southern Hemisphere, particularly in Africa, South America, and Australia. Thus, he was able to test the theory against his vast personal experience. His initial salvo in the debate came in 1927 in a Carnegie Institution of Washington publication, "A Geological Comparison of South America with South Africa." He followed this in 1937 with a book, *Our Wandering Continents,* which contained enough evidence to overwhelm believer and unbeliever alike, but which was so disputatious that it offended many in the staid scientific community.

THE SLINGS AND ARROWS of outrageous critics clearly hurt Wegener. The more his theory was criticized, however, the harder he worked to gather evidence to support it. Wegener was a tenacious man. He knew not to let adversity discourage his pursuit of the truth. The philosophy that enabled Wegener to carry on despite his tribulations is revealed in a letter he wrote to Georgi while they were preparing for his final expedition to Greenland. "Whatever happens, the cause must not suffer in any way," Wegener wrote. "It is our sacred trust, it binds us together, it must go on under all circumstances, even with the greatest sacrifices."

In 1929 Wegener published the fourth edition of *The Origin of Continents and Oceans.* Because of the difficulty in keeping up with the literature, he had planned for it to be the final edition, as it was—but for different reasons than he had envisioned. After his death on the Greenland ice cap in 1930 only a few devotees kept the faith. Despite the efforts of highly regarded scientists like Holmes and du Toit, those who believed in continental drift were often regarded as crackpots. Most respectable earth scientists, smug in their knowledge of geology derived from dry-land surfaces, dismissed Wegener's theory as beneath contempt. Few opponents deigned to give it a fair test.

Nevertheless, Wegener and his adherents, even though greatly outnumbered, managed to stalemate their critics. Thus Wegener's theory of continental drift did not fade into historical obscurity, as did the ideas of Taylor and others. The impasse was not to be broken for decades. A little more than thirty years after Wegener's death, John Tuzo Wilson, a Canadian geophysicist, explained why the controversy had not been resolved.

"The question whether the continents have been fixed in approximately the same relative position since their creation, or whether they have moved has been debated for fifty years," Wilson wrote. "Perhaps the reason that this has never been settled is that much more is known about the continents than about the ocean floors, where the decisive evidence probably lies."

The Pathfinder

It was a common opinion, derived chiefly from a supposed physical relation, that the depths of the sea are about equal to the heights of the mountains; but this conjecture was, at best, only a speculation. Though plausible it did not satisfy. There were, in the depths of the sea, untold wonders and inexplicable mysteries. Therefore the contemplative mariner, as in mid-ocean he looked down upon its gentle bosom, continued to experience sentiments akin to those which fill the mind of the devout astronomer when, in the stillness of the night, he looks out upon the stars, and wonders.

Matthew Fontaine Maury, superintendent of the United States Naval Observatory in Washington, D.C., from 1844 to 1861, thus summed up in his book *The Physical Geography of the Sea* what was known about the ocean depths in the middle of the nineteenth century—that is, not much. He was determined to reduce the ignorance.

MAURY DID NOT have much experience with the oceans while growing up. Spotsylvania County, Virginia—where on January 14, 1806, Maury was born—was more than one hundred miles inland from the Atlantic coast. His parents, Richard and Diana Maury, took the family even farther away from the sea, moving across the Appalachians into Williamson County, Tennessee. Maury's father wanted him to become a farmer. But the young Maury had other plans. He wanted to learn about science and math, and viewed a career in the Navy as the best way to fulfill his ambitions. His older brother, John, had obtained an appointment as midshipman in 1809.

Maury had begged his father to send him to nearby Harpeth Academy so he could get an education. His father initially refused, saying he needed the boy on the farm, but fate—in the form of a forty-five-foot fall from a tree—intervened. Maury survived, but the injuries were so severe that a doctor advised his father that the boy, once fully recovered, should not attempt farm work for a long time, if ever. In 1818, as soon as Maury was able to ride the six miles to the school twice a day, Richard Maury allowed his son to enroll. Harpeth Academy provided Maury with the only formal education he would ever receive.

Maury considered entering West Point in 1823 to further his education, but in July of that year the family learned that his brother John had died of yellow fever while at sea. Richard Maury did not want to lose another son to the military. Maury thus abandoned hopes of attending West Point. However, in 1825 he secretly appealed to Congressman Samuel Houston of Tennessee to help him obtain a commission as midshipman. The future president of the Republic of Texas followed through, and the nineteen-year-old Maury learned that he had been appointed an acting midshipman on February 1.

Maury's ancestors had played a prominent role throughout Virginia's history. Because of his background he was assigned to the frigate *Brandywine,* which was to take America's old ally, the Marquis de Lafayette, back to France after a visit to the United States. While on board the frigate, Maury's education was limited to lectures on navigation. He found the text, *American Practical Navigator* by Nathaniel Bowditch, wanting. He knew there was much more to learn.

The *Brandywine* returned to the United States via New York, and Midshipman Maury was granted a short leave in which he visited relatives and friends in Virginia. When the leave ended, he returned to New York and the *Brandywine,* which was being readied for a planned three-year voyage to relieve the frigate *United States,* on patrol in the Pacific. The *Brandywine,* accompanied by the sloop of war *Vincennes,* sailed over the bar at the mouth of New York Harbor on September 3, 1826.

As the ships sailed toward South America, Maury began to appreciate the wonders of the night sky—making his time on night watch more interesting, and fueling his desire to learn. After a tense stop in Rio de Janeiro—Brazil and Argentina were at the time fighting for control of Uruguay (and the Uruguayans wanted to be rid of both!)—the *Brandywine* and *Vincennes* pressed on toward the southern tip of the continent. They got their first taste of the fierce storms of the Southern Ocean on November 30 while north of the Falkland Islands and later endured a seventeen-day passage around Cape Horn. The challenge of navigation under such conditions further inspired Maury to master the subject. The two ships reached the port of Valparaíso, Chile, on December 26, where they were joined by the *United States* on January 9, 1827. The men of the *Brandywine* and *Vincennes* assisted in badly needed maintenance work on the *United States* and an American Navy brig, the *Columbia.* Finally, on January 24, the *Brandywine, United States, Vincennes,* and the schooner *Dolphin* departed for Callao, Peru, where they arrived on February 9. One month later Maury was transferred to the *Vincennes,* which patrolled the coast of South America for two years.

Maury spent his time in South America fruitfully. He did not learn much from the *Vincennes*'s schoolmaster but took advantage of the ship's library to study navigation and mathematics. He learned Spanish from the locals who invited him and other young officers to their homes, and spent much of his free time on land traveling and studying the region's geography. He was a keen observer of the region's political upheavals as well—and there was much to observe as Simón Bolívar's brief confederation, Gran Colombia and Peru, began to fall apart.

By late June 1829, tensions had decreased to the point that the *Vincennes* could be released from patrol duty. The navy had big plans for the ship. It was ordered to cross the Pacific and Indian Oceans and to return to New York via the Cape of Good Hope, thus becoming the first United States warship to circumnavigate the globe. The significance of the adventure was apparent to all Americans present at Callao. The crews of the other three American warships cheered on July 4 as the *Vincennes,* under the command of Master Commandant William Bolton Finch, followed the sunset out of the harbor.

The Reverend Charles S. Stewart, chaplain of the *Vincennes*—and former missionary to the Sandwich Islands (now Hawaii)—was impressed by Maury's desire to learn and made sure to include him on every excursion for which Maury could secure permission to leave the ship. The first island the *Vincennes* called upon was Nuku Hiva of the Washington Group (now the Marquesas). Maury was particularly interested in this island, for his brother John had been stranded there for almost two years until being rescued by Commodore David Porter in December 1813. Maury explored the territory, learned the culture and language of the native Happa people, and even met the chief who had befriended his brother. The *Vincennes* nearly met disaster on August 11, when the wind died and powerful currents nearly drove the ship, stern-first, into a cliff. The crew averted disaster by desperately thrusting spars out from the poop deck to keep the hull from striking rock. A breeze from the land eventually filled the topsails and nudged the ship out of the way of destruction.

After stops in Tahiti and the Sandwich Islands, the sloop made way for Macao, which it reached on New Year's Day 1830. Maury was selected to join a party of thirteen officers accompanying Finch to the restricted city of Canton to gather information on Chinese trade. After completing their mission, the men returned to the *Vincennes* on the eighteenth and departed four days later.

Over the next five and a half months the sloop made stops in the Philippines, Java, Cape Town, and St. Helena before returning to New York Harbor on June 8. Maury had been away from the United States for almost four years. He had seen a lot of the world, but had accomplished a lot more than mere sightseeing. He had learned much about foreign cultures, international relations, and geography and had gained valuable experience about practical matters of navigation, ship handling, and weather, winds, and currents.

Unfortunately, Maury didn't learn how to conceal his brilliance when judged by men of lesser intellect. He had devoted a considerable amount of time learning the science behind the art of navigation, but the members of the board of examination that he faced on March 3, 1831, were only interested in hearing him regurgitate the writings of Bowditch. His advancement was set back several years when the board asked him how to determine longitude at sea from a lunar observation. All the board wanted to hear was a memorized recapitulation of two methods set down in *American Practical Navigator.* Maury instead stepped up to a blackboard and worked out a solution using the principles of spherical trigonometry. The examiners, completely baffled by facts and reason, simply declared Maury mistaken and suggested that he return to the sea and "learn his business." He managed to pass the examination anyway, but he was ranked twenty-seventh in a class of forty when he probably should have finished first.

Maury's next assignment was as sailing master of the *Falmouth,* sister ship to the *Vincennes,* which was to depart for patrol duty off the west coast of South America. He reported for duty on June 11, 1831. As sailing master he was responsible for maintaining the seaworthiness of the ship, managing the supplies, and—more important—navigation. He kept detailed notes on the latitude, longitude, heading, distance traveled, tides, currents, wind direction, weather, astronomical observations, compass variations, and unique coastal features useful to navigators. He also examined existing charts for inaccuracies.

The *Falmouth,* accompanied by the HBMS *Volage,* began to round Cape Horn in October 1831. The two ships, caught in a gale off the cape, split up. The *Volage* kept to an inshore course but was beaten back twice before getting around. It sustained damage, however, and had to put into Talcahuano, Chile, for repairs. The *Falmouth* tacked down to 62° 5′ S, caught favorable winds, and made the west coast in fine order. Maury had passed his first real test as sailing master.

Prior to the voyage, Maury had searched for some kind of sailing guide to assist in his planning, but was unsuccessful. He knew the benefit of accurate information on sailing conditions and directions. He resolved to begin providing that information and, in his small stateroom, began to write what would become his first scientific paper, "On the Navigation of Cape Horn," which he submitted to the *American Journal of Science and Arts.* It was published in the July 1834 issue. While on patrol Maury spent as much time as possible surveying the west coast of South America. He also began to plan a new book on navigation for midshipmen.

Following a brief stint as executive officer for the schooner *Dolphin,* Maury was transferred to the frigate *Potomac* in the fall of 1833. The *Potomac,* sister ship to the *Brandywine,* was about to depart for the United States. Maury was eager to return, primarily to be reunited with his fiancée, Ann Hull Herndon, but also to find a way to supplement his meager navy salary so that the two could finally afford to get married.

On February 9, 1834, the *Potomac* sailed out of Valparaíso for the return voyage to the United States. On March 10, while rounding the Horn, the frigate began to be whipped by strong gales and had to weave its way through a field of massive icebergs. Despite the efforts of the crew, the ship struck a berg on the afternoon of the thirteenth. Fortunately, however, the ship was not seriously damaged. Dense fog the following day reduced visibility to zero. However, the frigate avoided further mishap and emerged into fair weather on the sixteenth.

The *Potomac* arrived at the Charlestown Navy Yard in Boston, Massachusetts, on May 24. Three days later Maury received a three-month leave of absence. He hurried to Fredericksburg, Virginia, to reunite with Ann. They were married on July 15. After the honeymoon ended he began studying geology, mineralogy, and astronomy as well as working feverishly on his planned textbook on navigation. He had plenty of time to write, as he was "awaiting orders" from the navy. On April 30, 1835, he finished his book. He submitted the manuscript, "A New Theoretical and Practical Treatise on Navigation," to Key and Biddle, a publishing firm in Philadelphia. The manuscript was accepted.

When the book was published in 1836, it was reviewed in the June issue of *Southern Literary Messenger.* The reviewer was Edgar Allan Poe. Impressed by the book's style as well as its substance, the not yet famous master wrote:

> The spirit of literary improvement has been awakened among the officers of our gallant Navy. We are pleased to see that science is also gaining votaries from its rank. Hitherto, how little have they improved the golden opportunities of knowledge which their distant voyages held forth and how little have they enjoyed the rich banquet which nature spreads for them in every climate they visit! But this time is coming when, imbued with a taste for science and a spirit of research, they will become ardent explorers of the regions in which they sojourn. Freighted with the knowledge which observation can only impart, and enriched with collections of objects precious to the student of nature, their return after the perils of a distant voyage will then be doubly joyful. The enthusiast in science will anxiously await their coming, and add his cordial welcome to the warm greetings of relatives and friends.

Praise came from Maury's navy colleagues as well. Eventually, on September 4, 1844, Navy Secretary John Y. Mason ordered that the book replace *American Practical Navigator* as the standard text for midshipmen.

CONGRESS on May 18, 1836, passed an act authorizing the navy to send an expedition to explore and survey the Southern Ocean. Political friction, however, primarily between President Andrew

Jackson and his secretary of the navy, Mahlon Dickerson, mired preparations for the voyage. Jackson appointed Captain Thomas ap Catesby Jones to command the expedition, but Dickerson, who was in no way an admirer of Jackson, obstructed Jones in every way. Maury, who had been commissioned a lieutenant on June 10, was likewise caught up in the fray. Because of Maury's burgeoning scientific reputation, Jones had sought his participation. Before long, Maury became embroiled in a brouhaha with Lieutenant Charles Wilkes over the custody of navigational and scientific instruments requested by Jones for the expedition. Wilkes, a favorite of Dickerson, hoarded many of the instruments in New York, but Maury eventually managed to recover most of them.

After months of battling with Dickerson, Jones asked—for a third time—to be relieved of command. His request was granted on December 6. When other, more senior, officers declined to be mistreated by Dickerson, command fell to Wilkes. Maury refused to work with him and requested to be reassigned. He was relieved (in more ways than one) on June 9, 1838.

Maury vented his frustrations over the incident through his pen, writing a series of articles—under the pseudonyms of Harry Bluff and Will Watch—that deplored the state of the navy and suggested badly needed reforms. The articles appeared in the Richmond (Va.) *Whig and Public Advertiser* and attracted great interest.

Early in 1839 Maury assisted in a survey of harbors along the coast of the Carolinas and Georgia. Later he was assigned to the brig *Consort,* which was being readied for service in New York. He took advantage of a lengthy leave to visit his aging parents in Tennessee. In October he began his return to New York and the waiting ship. He never made it. On the night of the seventeenth, prior to leaving Lancaster, Ohio, the gentlemanly officer gave his seat inside the overloaded stagecoach to a woman passenger. He crawled on top of the coach with the driver and two other men. A little past midnight the driver, busily engaged in conversation with the other two men on top, lost control on a detour. The stagecoach overturned, throwing Maury through the air. He was severely injured with multiple fractures of his right leg. Maury didn't know it yet, but his seagoing career was over.

"HARRY BLUFF" returned to the task of reforming the navy while Maury recovered from his injuries. In a series of articles in the *Southern Literary Messenger* in 1840 and 1841, he called for an increase in the size of the navy as well as for a series of coastal fortifications to protect the nation from attack. He demanded reform in the way that the navy educated its young officers, calling for an academy similar to the army's West Point. He also demanded that midshipmen be schooled in a broader array of subjects. Early in 1841 he suggested administration of the navy be reorganized from a dysfunctional board of politically connected commissioners to a series of bureaus with divided responsibilities. The current commissioners fought back, but Maury, who had reached a national audience with his articles, had drawn blood—a lot of it. Reform was on the way. And Maury, along with the science of oceanography, was to benefit.

IN NOVEMBER 1841, shortly after learning that he was found unfit for sea duty on board a man-of-war, Maury was elected a corresponding member of the National Institution for the Promotion of Science (which later merged with the American Association of Geologists and Naturalists to form the American Association for the Advancement of Science). Members included Joseph Henry, first secretary of the Smithsonian Institution, and Alexander Dallas Bache, who later became director of the United States Coast Survey (and Maury's rival). The organization's first bulletin included a proposal by Maury to systematically study the bottom of the sea.

Maury was appointed superintendent of the navy's Depot of Charts and Instruments in June 1842. He reported for his new duties on July 4. It was an auspicious time, for Congress was debating two bills of interest. One, to reorganize the navy, was inspired by his articles in the *Whig and Public Advertiser* and the *Southern Literary Messenger.* The other was to appropriate funds to build a badly needed national observatory to be run by the navy. The observatory would also house the Depot of Charts and Instruments. The members of the House of Representatives and the Senate, in a hurry to adjourn the session and get home, passed both bills on the same day, August 31.

On October 1, 1844, shortly after declaring Maury's *Navigation* the standard text for midshipmen, Secretary Mason appointed Maury superintendent of the newly built Naval Observatory.

WHEN MAURY TOOK OVER the Depot of Charts and Instruments, the United States was dependent on other nations, such as England, for navigational charts—even for American waters. Maury, who had realized the need for accurate information while sailing master of the *Falmouth,* found the situation deplorable. He also realized that he had a treasure trove of information in the archive of the ships' logs maintained by the navy at the depot. Maury and his staff began organizing the logs by region, then began the tedious process of extracting and correlating the information contained within. His goal was to produce a series of charts that would advise mariners of the average conditions they would likely face and suggest the best course to take in a given region at a given time of year.

Maury did not stop at purely historical research. He also prepared a set of logs to be sent to cooperative mariners. While at sea, the mariners were to provide data on latitude, longitude, the direction and speed of currents and winds, barometric pressure, air and water temperature, cloud conditions, precipitation duration, and magnetic variation. A form designed for use aboard men-of-war included entries for dry- and wet-bulb temperatures (for determination of relative humidity), sea conditions, and specific gravity of seawater. When a ship returned to port, the captain was to send the forms to the depot for analysis. Those who completed the survey and sent in the logs were entitled to copies of the newly developed charts. The first of the wind and current charts envisioned by Maury was published in the fall of 1847. It was, ironically, drawn by Lieutenant William B. Whiting, who had been ridiculed by fellow *Falmouth* crew members fifteen years earlier when he suggested that Maury would later achieve great things.

Maury was confident that his charts would allow mariners to make faster, safer transits of the seas. The depot's Wind and Current Charts and accompanying Sailing Directions arrived just in time: A mass of humanity was shortly to dash off to California following the discovery of gold at Sutter's Fort in 1849. Captains who followed Maury's instructions from New York to San Francisco cut forty to forty-four days off their sailing time.

The charts further proved their utility at the end of 1853 when the *San Francisco*, carrying soldiers from New York to California, was dismasted by a powerful Christmas Eve storm while crossing the Gulf Stream three hundred miles from Sandy Hook, New Jersey. Wholly at the mercy of the wind and waves, the *San Francisco* was sighted by other ships that were unable to reach it because of the high seas. One, the brig *Napoleon*, sighted the *San Francisco* on the twenty-sixth and hurried into New York to give the location. The secretary of the navy, ready to send two revenue cutters out to rescue the passengers and crew of the stricken ship, asked Maury to predict where it would be found. He studied his charts, carefully made and double-checked calculations, and wrote detailed instructions for the cutters to follow. By the time the cutters arrived at the *San Francisco*'s location, three other ships, *Kilby*, *Three Bells*, and *Antarctic*, had taken off the survivors (179 people had been washed overboard). The rescue occurred precisely at the location Maury had predicted!

AT ONE EARLY stage of the charting project, Maury had sought ships to conduct his own research of winds, currents, and sailing routes. The Gold Rush prompted Congress to authorize three small navy ships for Naval Observatory research, but no operating funds were provided. Maury was happy to get the use of one, the schooner *Taney*, commanded by Lieutenant Joseph C. Walsh. In October 1849 the unstable ship (now made top-heavy by the addition of a load of research equipment on deck) set out from New York bound for the Canary Islands. Initially the weather was rough, but in November it calmed enough for the men to get to their scientific work. Every thirty or so miles they stopped to measure the water temperature on the surface and at depth. Every two hundred miles they attempted to sound (measure) the depths. The sounding device consisted of a large reel, supported on a frame, filled with miles of steel wire with a weight attached to the end. The wire broke on the first attempt, with almost six miles played out. On one occasion the reel broke loose as sailors were trying to remove it from the frame. The sailing master's collarbone was broken in the excitement, and some minor damage was done on deck. Some equipment was destroyed and more lost as cables broke, leaving Walsh with only a thermometer and a few sounding lines by the time the *Taney* reached the Canaries. The leaky schooner underwent emergency repairs in the Cape Verde Islands and subsequently raced back across the Atlantic, accompanied by the brig *Porpoise*, to St. Thomas in the Virgin Islands for more extensive repairs. In May 1850 the *Taney* left St. Thomas on a zigzag course back to New York. Upon arrival, Walsh wasted no time in pronouncing the ship unseaworthy. On the whole, the cruise was disappointing, but Walsh's persistence resulted in the collection of valuable data.

Maury next obtained the services of the brig *Dolphin* (the schooner of that name had been sold in 1835) for three cruises. The ship was originally outfitted with the same type of equipment as the *Taney*, and its crew collected the same kinds of data on the first two cruises. Lieutenant Samuel Phillips Lee, who led the first cruise from 1851 to 1852, charted a new navigational hazard, the unmarked Rocas Reef, off the coast of Brazil's Cape São Roque, but found that a number of other reported dangers did not exist. Lieutenant Otway H. Berryman led the second cruise from 1852 to 1853. Berryman's main goal was to make deep-sea soundings across the Atlantic.

The sounding device performed better on the first two *Dolphin* cruises than it had on the *Taney,* but Maury wasn't completely satisfied. He wanted a device that could also bring up a sample of the bottom, but without becoming so heavy that the line would snap on the way up.

For millennia, humans had measured the depths of water with some combination of line (for the actual measurement) and weight (to carry the line to the bottom). In shallow areas, such as lakes, rivers, some inland seas, and portions of the continental shelves, sounding was a fairly routine matter. But measuring depth proved to be a challenge in the deep sea. Sounding lines, whether made of rope, metal wire, or cable, were prone to snap under the tremendous weight—not only of the sinker, but of thousands of feet of the line itself hanging over the side. If the sinker was not sufficiently heavy, or if it had broken off, the line could play out for thousands and thousands of feet without ever hitting bottom. The problem wasn't that the bottom was missing but that deep currents caused the line to drift below the surface. Even when the bottom was reached, the weight of the line alone could continue pulling it from the reel. Whether or not the line behaved on its descent to the bottom, a method was needed to prove that the bottom had actually been reached.

A young midshipman working at the observatory, James Mercer Brooke, devised a solution to Maury's problems. His device consisted of a sampling tube attached to the end of the line. The sampling tube passed through the center of a weight—usually a cannonball with a hole drilled through it—which was fastened to the line just above the tube. The momentum of the tube and weight drove them into the bottom sediments on impact. The weight would automatically detach from the line, and the line and bottom sample could then be retrieved.

The device got its first test on the third cruise of the *Dolphin,* again commanded by Berryman. On July 7, 1853, Midshipman

J. G. Mitchell took the device out on a small boat and let it loose at 1:20 P.M. Six hours later, Mitchell successfully retrieved the sample tube. The recorded depth was 2,000 fathoms, or 12,000 feet. The ooze that the device recovered from that and other soundings was filled with microscopic foraminifera and diatoms. Foraminifera and diatoms are single-cell organisms whose shells, in the case of forams, or cell walls, in the case of diatoms, leave telltale remains in marine sediments (and fossil rocks). Their presence in the deep-sea sediments surprised both Maury and West Point professor Jacob Whitman Bailey, who analyzed the samples.

Working with the sounding data he had received by 1852, Maury prepared a profile of the Atlantic Ocean floor between the Yucatán Peninsula and the Cape Verde Islands. After the completion of the *Dolphin*'s third cruise he had two hundred soundings from the North Atlantic. He then began preparing a map of the North Atlantic seafloor, which was published in 1854. The *Bathymetric Map of the North Atlantic Basin with Contour Lines Drawn in at 1,000, 2,000, 3,000 and 4,000 Fathoms* showed a broad plateau in the center of the ocean. Maury named it the Dolphin Rise. The map instantly became popular, especially with Cyrus West Field and his associates, who were planning to lay a telegraph cable across the ocean.

What Maury had called the Dolphin Rise turned out to run the length of the Atlantic seafloor, north to south. The undersea mountain range was to perplex scientists for more than one hundred years.

IN 1853 the publishers of Maury's navigation textbook suggested that he write a book about the ocean for the general reader before a professional writer did so and reaped the fame and fortune that Maury deserved. They further suggested that he call it *The Physical Geography of the Sea*. They even suggested that a firm larger than themselves that could achieve a much wider distribution—a firm such as Harper and Brothers in New York—publish the book. Maury appreciated the suggestion, Harper and Brothers liked the idea, and Maury began writing in the spring of 1854 in his spare time. With ample help from his family, he completed the manuscript on June 20, 1854, and it was published early in 1855.

The Physical Geography of the Sea was a tour de force. Maury's observations were astute, his explanations less so—due to his largely haphazard education—but his writing was moving. He began the book with a chapter on the Gulf Stream. These are his opening words: "There is a river in the ocean. In the severest droughts it never fails, and in the mightiest floods it never overflows. Its banks and its bottom are of cold water, while its current is of warm. The Gulf of Mexico is its fountain, and its mouth is in the Arctic Seas. It is the Gulf Stream."

Maury's treatise touched upon many topics: the atmosphere and weather conditions, currents, chemistry—and the depths. "Could the waters of the Atlantic be drawn off, so as to expose to view this great sea-gash, which separates continents, and extends from the Arctic to the Antarctic, it would present a scene the most rugged, grand, and imposing," he wrote of the Atlantic Basin. "The very ribs of the solid earth, with the foundations of the sea, would be brought to light, and we should have presented to us at one view, in the empty cradle of the ocean, 'a thousand fearful wrecks,' with that dreadful array of dead men's skulls, great anchors, heaps of pearl and inestimable stones, which, in the dreamer's eye, lie scattered on the bottom of the sea, making it hideous with the sights of ugly death."

THE BOOK was hugely successful. It was reprinted repeatedly in the United States and Great Britain and translated into Dutch, German, French, and Italian. But Maury did not have long to enjoy its success. Sectional tensions between the North and South were tearing the United States apart. Maury encouraged some kind of reconciliation, believing that the problems the nation faced could be solved only within the Union. He had hoped that Virginia would remain a part of the nation and realized that war would likely result if the commonwealth seceded. His hopes for peace dimmed after Confederate forces attacked Fort Sumter and President Abraham Lincoln called upon loyal states to provide troops to bring the rebels back into the Union early in 1861. When word reached Washington, D.C., on April 19 that the Virginia Convention had voted to secede, Maury decided to follow his home state.

The next day was his last as an officer in the United States Navy. He took care of some correspondence, updated some files, sent some navigational instruments to the New York Navy Yard, and cleared his desk. That afternoon, at three o'clock, he turned over the public property in his possession, as well as his officer's sword, to Whiting. Maury then asked his secretary, Thomas Harrison, to write his resignation from the navy. Harrison, overcome with grief, could not. Maury wrote the resignation letter himself, signed it, and had it sent to Lincoln. Dressed in civilian clothes, he walked out the door of the observatory. Maury, the "Pathfinder of the Seas," also walked away from a storied career in oceanography.

Ninety years would pass before the navy again played such an important role in ocean science.

The Abyss

Matthew Fontaine Maury's 1841 proposal for the study of the deep sea grew primarily out of Maury's own desire to learn more about the depths. Ten years later, however, a group of men *had* to know more. The men, marines of the technological revolution spawned by Samuel Finley Breese Morse's invention of the telegraph, took up one of the most challenging tasks of the age—laying telegraph cables across the ocean floors.

Frederick Newton Gisborne, an English engineer who in 1852 had built an underwater telegraph link between New Brunswick and Prince Edward Island, was trying to do the same between Newfoundland and Cape Breton Island, Nova Scotia, in the fall of 1853 when his investors balked. Gisborne traveled to New York City to secure new financial backing early in 1854 and was introduced to Cyrus West Field.

Field, an American who had retired from the paper manufacturing business in 1853 after making his fortune, immediately saw the potential for a much longer cable. He wasn't interested in telegraphic connections between the North American mainland and islands off its coast. His goal was to lay a cable across the body of water separating Europe and North America—the North Atlantic. As Field began pondering the feasibility of such a project, he immediately consulted two people: Morse, to ask if the telegraph could transmit across such a vast distance; and Maury, to ask if the ocean would prove too great an obstacle to cross. Morse assured Field that the telegraph could transmit across such a great distance. Maury assured Field that the ocean could be crossed and studied his soundings and seafloor samples in search of the best route between the continents.

Field, assisted by his brother, David Dudley Field—a noted legal reformer—and by Peter Cooper—who had designed and built the *Tom Thumb,* America's first steam locomotive—organized the New York, Newfoundland, and London Telegraph Company. Field also obtained assistance from the governments of Newfoundland, the United States, and Great Britain. The Newfoundland government granted exclusive rights to land submarine telegraph cables on the coast of the province, while the Americans and British offered ships to conduct surveys and lay cable.

The United States initially responded in 1856 by dispatching the *Arctic,* commanded by Otway H. Berryman, who had earlier conducted several surveys for Maury. The expedition should have been under Maury's control—it was in offshore waters, which were the jurisdiction of the recently renamed Naval Observatory and Hydrographical Office—but, through bureaucratic skulduggery on the part of Alexander Dallas Bache and others, the ship was sent out under the authority of the Coast Survey, whose mandate was limited to relatively shallow, inshore waters. (Bache, Coast Survey superintendent, and many other formally trained scientists in America resented Maury's amateur status—as well as his success. Bache's hostility is ironic in that his great-grandfather—one of America's most famous scientists and inventors—was not formally trained in science either. His great-grandfather was Benjamin Franklin.)

Berryman, who had been so successful under Maury's command, ignored the lessons he had previously learned in deep-water surveys and used Coast Survey techniques and equipment. The sounding data from the *Arctic* expedition were all but useless. Berryman's soundings on the *Arctic* differed from those collected nearby on the *Dolphin*. In addition, soundings collected on *Arctic*'s initial eastward traverse from Newfoundland to Ireland differed from those obtained on the return traverse back to Newfoundland.

A huge brouhaha erupted in the United States when Maury questioned the accuracy of Berryman's soundings along the cable route. The most serious consequence of the controversy, however, was that data on the depth and location of the Dolphin Rise collected during the *Arctic* cruise differed from the findings of the earlier *Dolphin* expeditions.

Field had to obtain accurate information on the plateau before attempting to lay the cable. He turned to the British government for help, and the Admiralty dispatched the *Cyclops*, commanded by Lieutenant Joseph Dayman, to survey the route from Ireland to Newfoundland. Dayman, using a modified version of James Brooke's sounding apparatus, carefully conducted the survey early in 1857 and confirmed Maury's earlier work.

A combined fleet of American and British ships sailed from Valentia Island, Ireland, on August 7, 1857, laying cable as they went. The cable broke on August 11, 330 nautical miles out in two thousand fathoms of water. Crewmen could have attempted to splice the line, but after meeting to discuss the issue, the officers and engineers concluded they did not have enough cable to reach Newfoundland. The Atlantic Telegraph Company—successor to the New York, Newfoundland, and London Telegraph Company—decided to postpone further work until 1858.

The plan in 1857 had been to begin laying cable at either side of the ocean and to splice it in the middle. In 1858, however, the cable was to be spliced in the middle of the Atlantic first. The ships would then part, one group heading east to Ireland, the other west to Newfoundland, laying cable as they approached the shores. The British ship of the line *Agamemnon* and the American frigate *Niagara*, both laden with cable and accompanied by two other British warships, *Valorous* and *Gorgon*, set sail from Plymouth, England, on June 10. The ships battled furious storms en route to the midocean rendezvous and encountered a number of difficulties laying the cable, quickly losing three hundred miles of line due to repeated breaks. The ships returned to Queenstown (now Cobh), Ireland, and the board of directors of the Atlantic Telegraph Company discussed what to do next. Some wanted to abandon the cable effort altogether, but Field, backed by physicist William Thomson, won approval for one more attempt.

The four ships sailed again on July 17 and spliced the cable shortly after noon on the twenty-ninth. The *Niagara* anchored in Trinity Bay, Newfoundland, at almost the same hour that the *Agamemnon* reached Dowlas Bay on Valentia Island. The cables were connected to telegraph stations on land, and on August 16 Queen Victoria sent President James Buchanan the following message: "It is a triumph more glorious, because far more useful to mankind, than was ever won by conqueror on the field of battle."

Unfortunately, the cable broke a few weeks later, and engineers could not locate the breach. Even more unfortunate, however, was that the blood spilt on fields of battle in the United States prevented further attempts to link the continents until August 1866, when Field finally achieved—and was able to sustain—his goal.

AS LATE AS the 1850s scientists wholeheartedly embraced two misconceptions about the deep sea. The first was that seawater, like pure water, was densest at a temperature of thirty-nine degrees. This led to the belief that water temperatures at the bottom of the ocean were, of course, thirty-nine degrees. Scientists reasoned that because no winds disturbed the abyssal waters and because it was "impossible" for an influx of denser water to displace the water already there, no movement, and therefore no exchange of gases and food, was possible. This led to the second major misconception: that the "stagnant" abyssal waters were devoid of life. They formed, in other words, an azoic zone. Edward Forbes, an English scientist, had determined that this lifeless layer began at about three hundred fathoms.

Evidence to the contrary was amazingly ignored. For example, in 1818 a young James Clark Ross—who would receive a knighthood in 1844 after leading a famous expedition to Antarctica—accompanied his uncle, John Ross—himself knighted in 1834—on an expedition to Canada's Baffin Bay. On a dredge in one thousand fathoms of water the young Ross hauled up a greenish mud teeming with worms. Neither mud nor worms float exceptionally well, but no one saw fit to discard the beloved concept of the azoic zone.

James Clark Ross was not without blame, for he confirmed in his own mind the existence of a deep, immobile, thirty-nine-degree layer of water between the latitudes of forty and sixty degrees. His mistake was using unprotected thermometers to measure the temperature of water at depth. Water temperature decreases about two degrees for every 550-fathom increase in depth in the South Polar seas, down to an ultimate low temperature of about thirty-five degrees. However, water pressure on an unprotected thermometer "increases" the temperature reading by about two degrees for every 550-fathom increase in depth. Ross also hauled up a number of mysterious creatures from the deep on his Antarctic expedition, but still he clung to the azoic theory.

As more and more animals were discovered in ever deeper waters, the boundary of the azoic zone was adjusted, but the concept itself was not discarded. The azoic zone began to lose its vigor, however, as work progressed on undersea telegraph cables. Initial reports of animals clinging to cables retrieved from the bottom were dismissed as instances where animals attached themselves while the cables were being raised through surface waters, but doubters soon emerged from among the faithful. In 1857 the azoic zone was finally dealt the death blow it deserved when a cable that failed after three years of operation was retrieved from thousand-fathom waters off the coast of Sardinia. Animals clung to it, some having cemented themselves to it. They could not have done so in the time it took to raise the cable.

THE PUBLICATION OF Charles Darwin's *On the Origin of Species* in 1859 sparked a search for missing links in evolutionary lineages among organisms all over the world, including the deep seas. Michael Sars, a Norwegian naturalist and theologian, had assembled quite a collection of bottom-dwelling animals from the North Sea and elsewhere. He had recognized a number of "primitive" species in his collection, including some that had previously been known only as fossils. As more and more collections brought up "primitive" animals, the belief grew that the depths harbored a vast community of "living fossils."

Thomas Henry Huxley, Darwin's most ardent defender, became involved in marine studies in 1846 when he served as surgeon-naturalist aboard the Royal Navy's *Rattlesnake* expedition to South America. Huxley continued his research on marine life and in 1857 examined bottom sediment samples collected during the *Cyclops* survey of the route of the telegraphic cable. While looking at the sediments with a microscope, he noticed close similarities between the material the *Cyclops* brought up and the widespread chalk beds of England. He came to the conclusion that the deep-sea sediments eventually became chalk deposits through compaction and uplift.

In 1868 Huxley reexamined the sediments collected by the *Cyclops* in 1857. He was surprised to find the sediments covered by a thin layer of jelly in which granular clusters were imbedded. He concluded that the combination represented a mass of living protoplasm—a living thing—and called it *Bathybius baeckelii. Bathybius* was believed to have dined on nutrients contained within the sediments and was supposed to be a virtually inexhaustible food supply for bottom-dwelling animals!

None of these conjectures had the least bit of truth to them. But they inspired scientists to begin a thorough examination of the deep sea.

IN 1868 Charles Wyville Thomson, a Scot working as professor of natural sciences in Belfast, Ireland, was involved in a study of echinoderms along the Irish coast with William Benjamin Carpenter, an English physician—and vice president of the Royal Society of London with a special interest in zoology. The two began to assess what was then known about the depths and realized that the times called for fewer deductions and more facts—and that the only way to get the facts was by launching an expedition to study the deep seas. Thomson suggested that Carpenter seek support for such a voyage from the society as well as the Admiralty. The primary goals of the expedition Thomson proposed would be to learn what animals lived in the depths, where the animals were located, and to determine the relationships between current species and their fossil forebears.

The Admiralty and Royal Society endorsed the plan. The Admiralty lent the expedition what Thomson referred to as a "cranky little vessel," the small steamship *Lightning,* and outfitted it with a steam-driven winch for dredging duties. The ship, whose primary distinction was that it was the oldest paddle-wheel steamer in the Royal Navy, was at the time employed as a surveying vessel. The Royal Society provided materials for preservation of specimens collected during the voyage.

The *Lightning,* with Thomson and Carpenter aboard, sailed from the Isle of Lewis on the west coast of Scotland on August 11, 1868, and headed for the Faeroe Bank, about two hundred miles north of the British Isles. The decks were jammed with bags of extra coal. The two scientists set up house in cabins that let in seawater as well as sunlight.

The August weather was unsettled, keeping the collecting equipment idle most of the time. What little dredging and sounding Thomson and Carpenter could manage was in shallow waters, where they found nothing of note. The scientists were not able to get much work done in the deep waters they sought until the beginning of September. Their patience was rewarded. "Very wonderful creatures" were found in five hundred fathoms of water, including a crimson and orange starfish with many arms, sponges, sea urchins, and deep-sea scallops equipped with hundreds of eyes—a bizarre characteristic for an animal that usually never sees sunlight.

Thomson and Carpenter also discovered that the water temperature at the ocean floor was thirty-two degrees, seven degrees colder than the men thought possible. At a new location a few days later, the water temperature at five hundred fathoms was a balmy forty-eight degrees! The temperature of the water near the surface of both areas was similar, however. The bottom was covered in "Atlantic ooze" inhabited by fragile glass sponges, sea pens, corals, sea urchins, brittle stars, and crustaceans. Another round of storms chased the *Lightning* back to Scotland after six weeks at sea.

The weather had allowed dredging for only ten days—and the steamer was in deep water for only four of those—but the dredging results proved that an abundant and diverse array of species lived in the depths. Temperature data showed the depths to be much more complex than previously imagined. For one, the measurements proved that the supposed uniform, stagnant thirty-nine-degree layer of water did not exist. The temperature differences also indicated that deep currents were likely.

Thomson, in his 1873 book, *The Depths of the Sea,* wrote that "it had been shown that great masses of water at different temperatures are moving about, each in its particular course, maintaining a remarkable system of ocean circulation, and yet keeping so distinct from one another that an hour's sail may be sufficient to pass from the extreme of heat to the extreme of cold." Thomson deduced that the temperature differences must result from the existence of a ridge in the ocean floor that prevented mixing of water in the depths. Water temperature, as well as bottom characteristics, seemed to have a significant effect on what type of animals lived on the ocean floor.

The results of the *Lightning* cruise impressed the members of both the Royal Society and the Admiralty, which supplied two better ships, the *Porcupine* and *Shearwater,* for extensive cruises in the North Atlantic in 1869, 1870, and 1871. A wealth of new species was discovered, and abundant life was found even at depths greater than 2,400 fathoms.

SHORTLY AFTER the Civil War, the United States Coast Survey resumed research cruises of its own. In 1871 the survey announced plans for a deep-sea expedition, led by Louis Agassiz, around South America. The threat of competition unnerved the British. Carpenter, in a note in *Nature,* sounded the alarm. In the article, dripping with wounded imperial pride, he warned that the British, "having shown other nations the way to the treasures of knowledge which lie hid in the recesses of the ocean . . . we are falling from the van into the rear, and leaving our rivals to gather everything up. Is this creditable to the Power which claims to be Mistress of the Seas?"

Apparently many others in the Royal Society and Admiralty were concerned, too. They responded with an endeavor that was far more ambitious than anything the ex-colonial upstarts from North America—or anyone else, for that matter—could muster. On December 7, 1872, a warship converted into a seagoing laboratory sailed out of Sheerness, England. That ship was named *Challenger.*

Challenger

Challenger, a 2,300-ton screw corvette, was built at Woolwich Dockyards, near London, in 1858. The three-masted square-rigger, two hundred feet long with a forty-foot beam, was equipped with a 1,200-horsepower steam engine and could carry a complement of about 240. In refitting the ship for the expedition, all but two of the guns were removed. The space freed up by the absence of weapons and ammunition was converted into laboratories, workrooms, and storage for equipment and specimens to be collected. An 18-horsepower steam winch was fitted on deck to provide power in support of seamen trying to haul up dredges and trawls from the deep. Hundreds of miles of hemp rope were stowed below deck to be used in sounding as well as equipment to measure temperature and currents and to obtain sediment samples from as deep as four thousand fathoms. For soundings the crew used a Baillie sounder, which was a modification of Brooke's famous design.

Captain George Strong Nares, born in Aberdeen, Scotland, was in command of about 20 officers, more than 200 crewmen, and a Newfoundland dog. Nares was familiar with scientific expeditions, having served on the *Shearwater* cruise. His second-in-command was Commander John MacLear. Lieutenant Thomas Henry Tizard, another veteran of the *Shearwater* expedition, served as navigation officer. The trio were up to the challenge of navigating the ship safely through the iceberg-ridden waters of Antarctica, the uncharted reefs of South Pacific atolls, and the furious winds of the Strait of Magellan. In addition to operating the ship, naval personnel were responsible for making meteorological and magnetic observations and handling the deep-sea sampling equipment.

Charles Wyville Thomson, who in 1870 had become professor of natural history at the University of Edinburgh, was in charge of five other scientists. Among them was John Murray, who had been born in Ontario, Canada, of Scottish parents. Murray was recommended—and more important, was available—when one scientist invited by Thomson had to back out shortly before *Challenger* sailed. Murray had not been known as a model student in the ancient halls of Edinburgh. He cared little for examinations and even less for completing degrees. He initially enrolled as a medical student but quickly grew weary of the lectures. Murray's intellect and interest in natural history, however, could not be tarnished by his disdain for studies. He managed to talk himself onto an 1868 Arctic cruise of the whaler *Jan Mayen* as a surgeon. During seven months in the far north he collected marine animals, conducted a bit of oceanographic research, and even visited the ship's namesake island in the Greenland Sea. He continued his oceanic researches off the coast of Scotland in subsequent summers; one year he discussed phosphorescence at sea with the physicist William Thomson on the Isle of Skye. He proved to be a worthy—and fortunate—addition to the staff.

Another Edinburgh scientist, John Young Buchanan, served as the expedition chemist and physicist. Rudolf von Willemoës-Suhm, a German who left a post at the University of Munich to join the expedition, and Henry Nottidge Moseley, a British scientist recently returned from a trip to Ceylon (now Sri Lanka), were the other two naturalists aboard. James John Wild, a Swiss artist, completed the scientific staff as artist and secretary. In addition to their studies of ocean life and characteristics, the scientists were to explore the lands they visited. Moseley, in particular, relished the chance to observe the flora, fauna, and humans along the distant shores.

The *Challenger*'s shakedown cruise to the Canary Islands was a stormy one. When the weather cleared, the crew practiced sampling techniques. They needed to. Hemp rope a little more than a half inch in diameter was selected for use as a sounding line because piano wire—which had become the material of choice—could not carry thermometers, density gauges, and other instruments the scientists wanted to use to measure the characteristics of the deep. The sounding line, however, broke three times on the way to the Canaries. The dredge rope broke once, and the dredge itself came up fouled on one occasion. The seamen quickly came to appreciate the tedium of the work. Dredging and trawling in deep water often took all day.

The original type of dredge used was soon scrapped for a beam trawl designed by Nares. It consisted of a wood beam with iron brackets at either end. A V-shaped net, fitted with lead weights to keep it on the bottom, was attached to the brackets. "Bread-bag stuff" lined the end of the net to keep small organisms from being lost.

Challenger left Tenerife on February 14, 1873, bound for St. Thomas in the Danish West Indies (now the U.S. Virgin Islands). The crew settled into what was to be their routine for the next three and one-half years. Every two hundred miles the sails were furled, the boilers were fired up, and steam raised. The ship was headed into the wind, and sampling gear was lowered over the side. *Challenger* had to remain still while sounding, but while dredging—or "drudging," as the seamen called it—the ship was allowed to drift, broadside to the wind, for a half hour or more before the dredge was retrieved. Moseley, for one, grew weary of the racket created by the steam winch. "From daybreak to night the winding-in engine was heard grinding away with a painful noise," he later wrote in his 1879 account of the expedition, *Notes by a Naturalist on the "Challenger."* Once the dredge was safely on deck, the sails were lowered and *Challenger* resumed its journey. While the ship was underway, the contents of the dredge were examined. Animals large enough to be seen were collected, while the sediment, by the shovelful, was washed through a set of nested sieves with increasingly finer mesh size so that smaller biological and geological treasures could be found.

St. Thomas was reached on March 16, and *Challenger* put in for a refit and to take on coal. Afterward the ship sailed north. The crew were dredging in waters near Bermuda when the dredge snagged on the bottom. The line broke, letting loose a heavy block on deck that struck and mortally wounded a young seaman, William Stokes. The next day, after Thomson assured a delegation of superstitious sailors that, contrary to traditional beliefs, Stokes's weighted corpse would sink to the bottom rather than float forever at some intermediate depth, the body was committed to the deep—3,875 fathoms, to be exact.

Over the next few months *Challenger* zigzagged across the Atlantic, stopping at Halifax, Nova Scotia, back again to Bermuda, and then east to the Azores and the Cape Verde Islands. After taking on coal and other supplies, the ship headed west to Brazil, stopping in the middle of the Atlantic just north of the equator at St. Peter and St. Paul Rocks. The islands, black at sea level, are encrusted in white elsewhere—the white being the dung of thousands of birds. The origin of the small group of islands was a mystery. They rise steeply out of the depths, so steeply in fact that it was impossible to find an anchorage. *Challenger* spent two days tied to a pinnacle along the shore. Some scientists had previously argued that the islands were the remnant of a landmass lost through submergence. A controversy erupted over whether the islands were of volcanic origin—Charles Darwin had said they were not, but others disagreed. The *Challenger* scientists could not solve the controversy. They left the mysterious islands fascinated, but with as many—or more—questions than before their arrival.

The corvette made a brief call at Fernando de Noronha, an island the Brazilians used as a penal colony, and reached Baía de

Todos os Santos, on the Brazilian mainland, on September 14, just in time to refill the ship's almost empty coal bunkers. Scientists and crew took advantage of the stop to go sightseeing. Their recreational activities, however, were cut short when one of the sailors came down with symptoms of yellow fever. *Challenger* left for colder—and healthier—climates, setting sail for the Tristan da Cunha Islands, in the middle of the South Atlantic at about 37° 15′ S.

The Tristan da Cunha Islands had been discovered—and quickly forgotten—about 1506 by the Portuguese. Perceived military necessity put an end to the islands' obscurity in 1816. The British, who exiled Napoleon to St. Helena, about 1,300 miles north of the islands, feared that they might be used as a base from which to liberate the fallen emperor. The Admiralty took possession of the islands and stationed a garrison there. Reason soon took hold of the British, however, who by 1817 had realized why the normally acquisitive Portuguese had left the islands alone. In the effort to evacuate the garrison, a Royal Navy sloop, the *Julia,* was wrecked on the north shore and took fifty-five of the complement of ninety-five down with the ship. Eventually most of the garrison was taken off, but a few men asked permission to stay behind on the main island, Tristan da Cunha. In time women were brought to the island, families were started, and a small but thriving colony was established. The industrious islanders raised cattle and sheep and grew fruit and vegetables—commodities highly sought after by the crews of ships who had spent a long time at sea. The prime purpose of the *Challenger*'s visit was to purchase beef and mutton, but the scientists managed to explore nearby Nightingale and Inaccessible Islands, where ill-mannered—and not in the least bit fearful—penguins in vast rookeries attacked their legs and feet. *Challenger* sailed for Cape Town, in the Province of Good Hope (now part of South Africa), on October 18, 1873, carrying two German brothers from Inaccessible Island whose dreams of riches in the fur seal business had failed to materialize and also carrying relieved scientists and seamen whose lower extremities needed time to recover from the onslaught of pugnacious birds.

By the time *Challenger* reached Cape Town, the ship had been at sea ten months. The next stage of the expedition was to be a long and dangerous passage through Antarctic waters on the way to Australia. The ship was to remain in port for seven weeks while preparing for the journey, and the crew was given liberal shore leave. More than a few were tempted by potential riches in the Kimberly diamond mines and did not bother to return. While Nares and his staff readied the ship, Thomson and the scientists were busy, sometimes exploring but mostly cataloging and packaging their wealth of plant and animal specimens. By the time their work was finished, they had filled sixty-four crates to be shipped back to England.

Challenger departed for Antarctic seas on December 17. Most of the crew, after almost a year at sea, had begun to grow weary of the dredging routine and were no longer interested in what was found. "At first, when the dredge came up, every man and boy in the ship who could possibly slip away, crowded round it, to see what had been fished up," Moseley wrote. "Gradually, as the novelty of the thing wore off, the crowd became smaller and smaller, until at last only the scientific staff, and perhaps one or two other officers besides the one on duty, awaited the arrival of the net on the dredging bridge, and as the same tedious animals kept appearing from the depths in all parts of the world, the ardour of even the scientific staff abated somewhat, and on some occasions the members were not all present at the critical moment, especially when this occurred in the middle of dinner-time, as it has an unfortunate propensity of doing."

Along the way to Antarctica, *Challenger* stopped to explore the Prince Edward, Kerguelen, and Heard Island groups. The fickle weather of the Southern Ocean upset the routine (and ruined an attempt to land on the Crozet Islands east of Prince Edward), but, when conditions permitted sampling, the results were rewarding—despite Moseley's complaint. Between the

Crozets and Kerguelen Islands, soundings revealed that the water was as much as 2,700 fathoms deep. Between Kerguelen and Heard Island, the ocean floor was much more shallow and variable, ranging from 100 to 425 fathoms. After the *Challenger* left Heard Island, it sailed due south.

The *Challenger* contingent got a rude welcome to Antarctic waters during a gale at one hour past midnight on February 8, 1874. A large wave slammed into the ship. The blow stove in two ports on the starboard side of the main deck, flooding—and carrying away everything in—the sick bay. Fortunately, no one was there at the time. Lookouts had been posted to watch for icebergs, and on the eleventh, the first one, 2,100 feet long and 220 feet high, was sighted. Also, as the ship moved south from the Heard Islands, the water deepened. The ship reached the edge of pack ice at three o'clock on the morning of the fourteenth. The temperature readings proved interesting. Water at the surface was 33 degrees. The temperature dropped to 29 degrees at 200 fathoms, then began to warm again. (Freshwater freezes at 32 degrees, but the freezing point decreases when substances are dissolved in it. Seawater, of course, is laden with dissolved salts.) Whatever life may have remained in the theory that the depths were no colder than 39 degrees was most certainly sapped by hypothermia at 200 fathoms. The ship crossed the Antarctic Circle on February 16.

Nares and his officers carefully steered *Challenger* through the wondrous world of ice. The blue-green sea thundered as the icebergs broke apart. The icebergs themselves were mostly white, but the recesses revealed a deep blue where compacted snow had been compressed into pure, crystalline ice. The sun added beautiful effects of its own. As the orb dipped low in the sky, the icebergs cast ominous shadows across the water, and the reds and golds of sunrise and sunset were reflected off the white of the clouds and ice.

The icebergs, though beautiful, were also dangerous. On the twenty-fourth, while the *Challenger* crew was dredging in 1,300 fathoms, the southeasterly wind blew into a gale. Falling snow greatly limited visibility. The ship, with close-reefed topsails, steamed into the lee of a big berg. Most of the time the force of the wind on the sails balanced the force of the screw, but for a moment the wind stopped and *Challenger* lurched forward into the iceberg, losing its jibboom. The crew managed to recover it. The corvette was difficult to control in the storm, and it drifted, broadside to the wind. As the ship was about to be driven into another iceberg, the wind let up enough to allow the crew to regain control and turn *Challenger* back into the wind. A more powerful storm battered the ship on the night of the twenty-sixth and twenty-seventh. But the winds slackened as dawn broke, and the corvette left the realm of the icebergs, bound for Australia.

Challenger reached Melbourne on March 17 and went round to Sydney to be refitted and reprovisioned two weeks later. Scientists and sailors alike took advantage of the break to travel on the continent, and more of the seamen deserted—this time heeding the call of Australian goldfields. Scientifically, the sojourn in Australia was disappointing.

The ship attempted to leave Sydney on June 8, but severe storms in the Tasman Sea forced it to return for shelter. Despite the fickle weather, *Challenger* managed to depart for good four days later, embarking on a somewhat circular course: east to New Zealand; north to Tonga; and west to Fiji, the New Hebrides, and eventually the Cape York Peninsula on the northeast coast of Australia. On August 19, in the Coral Sea west of the New Hebrides, the scientists found that the water temperature was the same (35.8 degrees) at 1,300 fathoms as it was at the bottom—2,650 fathoms. The pattern held until they reached the Great Barrier Reef: From about 1,300 fathoms to the ocean floor, the temperature was about 36 degrees. The scientists' concluded that a submarine ridge cut off the depths of the Coral Sea from colder waters.

After a sweltering week at Cape York, *Challenger* sailed through Torres Strait—between the cape and the southern coast of New Guinea—into the Arafura Sea. Repeated soundings revealed that the water was rarely more than forty fathoms deep. The two months from September 8, when *Challenger* left Cape York, until November 17, when the ship arrived in Hong Kong, were busy, with stops in the Aru, Kai, Banda, and Molucca Island chains (in the eastern part of the Malay Archipelago) and the Philippines.

Nares's time on the *Challenger,* to the regret of everyone on board, ended in Hong Kong. The captain had been selected to lead a two-ship expedition to the Arctic, and Nares, accompanied by his first lieutenant, Pelham Aldrich, departed for England.

(Unfortunately the Admiralty, which had an uncanny talent for getting its men killed in the Arctic—most notably the entire complement of officers and crew of the ships *Erebus* and *Terror* on the disastrous Franklin expedition of 1845 to 1848—had learned nothing from its failures and others' successes. Nares's hastily arranged expedition was dispatched without proper equipment and, more important, food. Nares's men trudged northward, with sledges too heavy for navigation on the perilous ice pack, and were weakened by scurvy, caused by a lack of vitamin C. Aldrich, with a sledge named *Challenger,* managed to set a record for the farthest north ever reached by humans, 83° 20′, but no one planted the British flag on the North Pole. Nares, upon his return from the Arctic in November 1876, was blamed for a "disaster" that wasn't. While men on previous expeditions had repeatedly suffered from "debilitation"—in other words, scurvy, even though it wasn't reported as such—his men were judged to be the first to have problems with the dreaded disease! Sir John Franklin had led all 129 men on his expedition, including himself, to their deaths; he was proclaimed a hero. Only four men on Nares's sledging parties died, but his record was forever stained. The Nares expedition reaped a wealth of scientific discoveries and claimed new territory for the British Empire. The board of inquiry, however, eager to shift blame from the Admiralty's institutional incompetence in polar matters, ignored what Nares achieved and condemned him for not serving his men enough lime juice.)

Captain Frank Turle Thomson, of the *Modeste,* a ship on the Royal Navy's China Station, replaced Nares as commander of the *Challenger.* The corvette, relieved of 129 cases and several casks of specimens collected since leaving Sydney, left Hong Kong on January 6, 1875. (Nares and his two ships, *Alert* and *Discovery,* sailed from Portsmouth, England, for the Arctic just four months later, on May 19.) The corvette skittered across the western Pacific: south again to the Philippines and the north shore of New Guinea, east to the Admiralty Islands, and then north to Japan.

On March 23 the ship stopped to sample at 11° 24′ N, 143° 16′ E, in waters between the Caroline and Mariana Islands. The sounding line, weighted with three hundredweight (about 336 pounds), was set overboard and took what seemed an excessively long time to pay out. The first reading was 4,575 fathoms. The officer on watch suspected that a strong current may have stripped the line. The crew tried again, this time with an extra hundredweight added, and the line was sent overboard again. The line stopped its descent at 4,475 fathoms—more than five miles—which was recorded as the official depth. The sampling tube, once retrieved, was filled with bottom sediments and mercury that had spilled out of thermometers crushed by the extreme pressure. The crew of *Challenger* had thus discovered the Mariana Trench, one of the deepest basins in the world. (Nearby the waters proved to be around 6,000 fathoms, or nearly seven miles, deep.)

The rest of the voyage to Japan was trying for the men. Moseley later wrote: "A most tedious voyage. The vastness of the expanse of water in the Pacific Ocean, in proportion to the area of the dry land, was pressed most strongly upon our attention." The lack of wind frayed the men's nerves, but it began to pick up as *Challenger* neared Japan. By April 11 the coast was in sight.

After a two-month stay in Japan, during which *Challenger* was dry-docked in a shipyard in Yokosuka, the crew set sail on June 16 bound for the North Pacific and Sandwich Islands (Hawaii). About a week out, sounding data confirmed the existence of another deep discovered the year before by an American ship, the *Tuscarora. Tuscarora* had reported a depth of 4,655 fathoms but recovered no sediment (the sounding later proved to be accurate). Farther to the south, *Challenger* recorded a depth of 4,000 fathoms. From the deep area off the coast of Japan, the bottom gradually rose as the ship neared the Sandwich Islands. Molokai was sighted on July 27. The ship made calls at Honolulu on Oahu and then at Hilo on Hawaii, where the scientists took advantage of a chance to see the massive volcano, Mauna Loa, and the gaping crater called Kilauea.

From Hawaii *Challenger* set a southerly course for Tahiti. Tragedy struck again as Willemoës-Suhm, the young German naturalist, contracted erysipelas—a bacterial infection of the skin—and died within days. His body was committed to 2,700 fathoms of water.

Few animals had been retrieved in the trawls since leaving Hawaii. But the dredges on September 11 began to bring up a number of puzzling rocks—manganese nodules—that seemingly littered the ocean floor. The origin of the rocks was a mystery. Some nodules contained teeth of fish long since extinct; in time ear bones of whales and dolphins were also found. The fact that fossil teeth and bones were found encased in the rocks proved that sediments accumulated slowly in the depths.

Tahiti was sighted on September 18. After a two-week stay in the area, *Challenger* set sail, first for the Juan Fernández Islands—whose most famous resident, the marooned sailor Alexander Selkirk, inspired Daniel Defoe's novel *Robinson Crusoe*—and then for Valparaíso, Chile, where the corvette would be readied for the fjord-lined passage through the Strait of Magellan. Valparaíso was reached on November 19; *Challenger* departed for the tip of South America on December 11, but foul weather slowed its voyage. The ship entered the inner passage of the strait on January 1, 1976. After a three-week transit of the narrow, winding, and treacherous waterway, *Challenger* reentered the Atlantic Ocean.

The corvette first sailed east to the Falkland Islands, then north to Montevideo, Uruguay. The toll of three years at sea was evident in the diaries of some of the scientists and crew. Moseley, for instance, had been a loquacious narrator of what he saw and did through most of the voyage. By the time he arrived in Montevideo, however, he had nothing to say, even though Uruguay was in the midst of violent political upheaval. For instance, the marble staircase of Montevideo's English Club was riddled with bullet holes where seven men had recently been killed. Moseley never mentioned the slayings.

Sampling continued, however, as *Challenger* sailed from Montevideo on February 25. The ship made its way east, then turned north in the middle of the South Atlantic north of Tristan da Cunha, making way for Ascension Island. The scientists encountered relatively shallow depths, about two thousand fathoms, along the way. Differences in temperature indicated that deep waters toward Africa were warmer than deep waters toward South America. The conclusion was obvious: an undersea ridge rose from the depths in the South Atlantic as well as the North.

The corvette stopped at Ascension Island and again at the Cape Verde Islands before working its last deep-sea station just south of the Azores. The exhausted group of scientists and sailors anchored off Spithead in southern England just before dark on May 24. Their travels had ended.

IN THE THREE and one-half years that *Challenger* was at sea, the scientists carried out the first systematic study of the world's oceans and firmly and finally exploded the notions that water at depth was a stagnant, motionless layer no colder than thirty-nine degrees; that life could not survive there (they found it even in the deepest depths); that the mysterious organism known as *Bathybius* made up the ooze on the ocean floor; and that sediment layers in the depths are destined to become uplifted to form sedimentary strata, like chalk beds, on the land surface. In all the oceans encountered, the expedition found currents that swirled at all levels, differing in temperature, direction, and speed of flow and in other characteristics such as salinity. The scientists found evidence that cold water originating at the poles could flow as far as the equator. While some "living fossils" were found on the ocean floor, a host of other organisms, some familiar, some heretofore unknown, were also discovered. A greater understanding of the origin of, and variability in, deep-sea sediments was gained. Unusual submarine topographic features, such as shallow ridges and deep trenches, were identified.

Charles Wyville Thomson was knighted upon the expedition's return to England. He wrote two volumes of what would become a fifty-volume report of the expedition, but died in 1882 at a relatively young fifty-two years of age, long before the project was completed. The death of the expedition's leader could have proved disastrous, but Thomson left a worthy successor, John Murray.

Murray, with his sharp eye, keen mind, and dogged perseverance, was probably the only man who could have filled the vacuum left by Thomson's death and saved the expedition. While *Challenger* was in the vicinity of Java, Murray had noticed bits of phosphate in the sampling nets and concluded that they must have originated on land. He later organized a search for the source, which was Christmas Island, south of Java. Great Britain annexed the island, and a company organized by Murray began mining the deposits. Murray earned a fortune.

He quietly used the money to support oceanographic research, and, when the British treasury balked at the cost of completing the *Challenger* expedition report, he took over, finishing the project at his own expense in 1895. When the treasury refused to pay for a *Challenger* medal, Murray again stepped up and saw to it that the award was distributed to everyone who had either served on the expedition or helped prepare the report. Murray's determination ensured that the legacy of the expedition would last for decades. Scientific publications based on the expedition's data appeared well into the twentieth century.

Thomson best summed up what *Challenger* accomplished. "The objects of the expedition have been fully and faithfully carried out," he wrote. "We always kept in view that to explore the conditions of the deep sea was the primary object of our mission, and throughout the voyage we took every possible opportunity of making a deep-sea observation. Between our departure from Sheerness on December 7th, 1872, and our arrival at Spithead on May 24th, 1876, we traversed a distance of 68,890 nautical miles, and at intervals as nearly uniform as possible we established 362 observing stations."

Probably no single expedition has revealed as many of the secrets of the deep.

Titanic Effect

Captain Edward James Smith was enjoying his final cruise, driving his ship, *Titanic,* west across the icy North Atlantic at a speedy twenty-two knots. The ship was a technological masterpiece, a gigantic liner 883 feet long, displacing 66,000 tons, with three screws that allowed the behemoth to slice through the sea at as much as twenty-five knots. It boasted the most majestic, luxurious accommodations of the age. Smith's record was perfect—he had never had a collision or any other major incident. Thus he was rewarded with the command of this ship, White Star Lines's jewel of the seas, on its maiden voyage from Southampton, England, to New York City.

Even though Smith had received warnings about icebergs in the area, the only precaution he took was to tell the lookouts to watch for ice. With all the attention on this much ballyhooed voyage—and with some of the most distinguished members of American and British society among his 2,207 passengers—Smith was not about to slow, or stop, his unsinkable ship. First impressions are too important.

And so are the last.

Shortly before 11:40 on this night, Sunday, April 14, 1912, lookout Frederick Fleet spied a dark shape ahead and informed the bridge. Thirty-seven seconds later, the ship started a slight turn to port. Fleet thought the big ship might miss, but as the bow slipped safely past the berg, many others on the *Titanic* felt a slight bump, a hiccup really, in the liner's momentum. Smith must have gravely pondered his retirement options in more than two miles of water as, at 2:20 A.M., the big ship's stern rose straight up and followed the rest of its torn body beneath the glassy surface on this calm, clear night.

More than 1,500 people died.

AT 9:00 P.M. on Christmas Eve, 1906, mariners in the Atlantic and Caribbean heard the familiar dot-and-dash pattern that in Morse code spells "CQ CQ CQ." The signal, a general call to all stations within range, was routine. As the mariners listened, however, they were shocked. A man's voice came on the air! Guglielmo Marconi's transmitters could send only dots and dashes—which sounded like dits and dahs—in the special language developed by Samuel Morse in 1838. The man behind the voice, Reginald Aubrey Fessenden, had for years wanted to achieve much, much more.

In 1900 Fessenden—a Canadian who had previously worked for Thomas Alva Edison and as a professor at Purdue University in Lafayette, Indiana, and the Western University of Pennsylvania (now the University of Pittsburgh)—had been developing a radiotelegraphy (wireless telegraph) system for the United States Weather Bureau. He became interested in transmitting the human voice and soon developed a way to transmit voice and other sounds by modifying the amplitude of a carrier signal while keeping the frequency constant. This method, amplitude modulation, laid the groundwork for the eventual development of AM and shortwave radio. Fessenden had set up an experimental station on Cobb Island, in the Potomac River in Maryland, and on December 23, 1900, he sent a message to a worker at a station one mile away: "One, two, three, four. Is it snowing where you are Mr. Thiessen? If it is, telegraph back and let me know." It was indeed snowing. Thiessen wired back.

Over the next six years Fessenden further developed two-way voice radio as well as a much more efficient transmitter for Morse Code that surpassed Marconi's work. Fessenden's signals could be heard as much as six thousand miles away. He beat Marconi by transmitting signals in both directions across the Atlantic—between test stations at Brant Rock, Massachusetts, and Machrihanish, Scotland—and he sent the first voice message across the ocean, also between Brant Rock and Machrihanish.

After calling all stations from his transmitter on Brant Rock on Christmas Eve, 1906, Fessenden announced the evening program. Then one of Edison's phonographs was cranked up, and the mariners, some of them hundreds of miles away, were serenaded by a recording of Handel's "Largo." Fessenden himself contributed to the entertainment, picking up his violin to sing and play "O Holy Night." Stage fright marred a Bible reading by his wife, Helen, and his secretary. Fessenden closed the broadcast with a hearty "Merry Christmas."

HAVING CONQUERED the airwaves, Fessenden in 1912 focused his attention on the problem of transmitting signals through water and began working for the Submarine Signal Company in Boston. The company was formed in 1901 to develop an underwater warning system to help ships avoid running aground. Early efforts were devoted to the design of submarine bells that could be heard at great distances. In time one hundred of the bells were deployed and more than 1,500 ships fitted with receivers.

Submarine Signal had experimented with using the bells to send Morse code signals through the water, but the effort failed. Fessenden began to design an oscillator that could produce a clear, powerful sound under water. After the *Titanic* disaster, Fessenden hoped the oscillator would be able to do more than send sound signals. He wanted it to be able to find icebergs.

On a snowy day in January 1914, Fessenden's oscillator was to be tested in the ocean for the first time. One tugboat, *Susie D,* carried the oscillator, suspended from a cargo boom. Another tug,

Neponset, carried receivers. A Morse code carrier was used to control the oscillator. *Susie D* anchored near the Boston Lightship while *Neponset* headed out to sea. The oscillator was turned on. Despite deteriorating weather, *Neponset* detected a clear signal as far as thirty-four miles away until the storm and darkness forced the tug to return.

Fessenden combined his oscillator, which would usually send out a "ping" sound, with a receiver that would pick up the echo as the signal "bounced" off an object and returned to the ship. The amount of time it took for the echo to return would be measured. An estimate of the distance to the object, whether iceberg, ocean floor, or—later—enemy ship, could then be determined by multiplying the time interval by the speed of sound in water—generally eight hundred fathoms per second—and dividing by two. (It was later found that the speed of sound value has to be corrected for differences in temperature, salinity, and pressure of the water. The adjustment, however, is quite small.)

Three months later the device Fessenden called an "Iceberg Detector and Echo Depth Sounder" was ready to be tested. The United States Coast Guard, recently placed in charge of the newly created International Ice Patrol—formed in response to the sinking of the *Titanic*—agreed to help test the device on a real iceberg. The iceberg detector was installed on the cutter *Miami,* which sailed soon after. The weather was horrible, most aboard were seasick, and the fog was so thick at times that the ship had to stop for fear of hitting—an iceberg!

After several days the fog lifted. Miraculously, an iceberg happened to be nearby. A lifeboat davit was used to lower the oscillator into the water. Return signals bounced off both the berg and the bottom. The results were summed up in a telegram sent to Boston that night. "Bottom one mile. Berg two miles," it said.

FESSENDEN CONTINUED to improve his design and eventually developed and patented the fathometer, which allows a ship to determine depth beneath its keel. He won the Scientific American Gold Medal in 1929 for the invention.

Others were working on the problem as well. Harvey Cornelius Hayes used a Fessenden oscillator as part of his Hayes Sonic Depth Finder. Hayes's device was the first capable of deep-water sounding. The Sonic Depth Finder was ready for testing in deep waters in 1922 and was first installed on the soon to be decommissioned battleship *Ohio* for a run between New York City and Annapolis, Maryland. The equipment was then transferred to the destroyer *Stewart* for a voyage across the Atlantic. The destroyer, with Hayes aboard, left Providence, Rhode Island, for a nine-day cruise to Gibraltar. During that short span, 900 soundings were obtained, and the Sonic Depth Finder worked fine even when the *Stewart* was slicing through the water at twenty-three knots. The results were spectacular, especially considering that the *Challenger,* in three and one-half years, obtained only a little more than 350 soundings.

(The *Stewart* was soon transferred to the Pacific and in World War II had probably the most unusual career of any American naval vessel. The ship was stationed at Tarakan in Borneo when the Japanese invaded the Dutch East Indies, now Indonesia, in 1942. In a battle off the coast of Bali early on the morning of February 20, the destroyer took a direct hit that flooded the steering-engine room, but it escaped. After reaching the Allied base at Surabaya, the *Stewart* was put in a floating dry dock at a private shipyard for repairs. The ship was badly braced, and when the dock was being raised it fell over on its port side. Retreating Allied forces destroyed the ship and the dock on March 2—or so they thought. From 1943 on, however, Allied air crews kept reporting a ship that looked a lot like a World War I–era American destroyer. They weren't seeing things. The Japanese had managed to raise, salvage, and recommission the *Stewart* as a patrol craft, PB-102. PB-102 is credited by some with sinking the submarine USS *Harder* on August 24, 1944, in a depth charge attack off Luzon in the Philippines. Among the casualties was its skipper,

Commander Samuel David Dealey, who was posthumously awarded the Medal of Honor for his exploits during the war. American forces recaptured the former *Stewart* in Kure, Japan, after the war ended. The ship was recommissioned but could get only its old number, DD-224, back, for the Navy had named another ship *Stewart,* believing DD-224 destroyed. Sailors nicknamed the ship "RAMP"—meaning "Recovered Allied Military Personnel." A prize crew returned RAMP to the United States. The unsentimental navy sank the ship during target practice off San Francisco in 1946.)

THE GERMAN NAVY was devastated by World War I. Its demise was not due to battle losses—the large fleet of dreadnoughts held its own at Jutland, and the submarine forces nearly brought the Allies to their knees—but came about by terms of peace when the bulk of the fleet was surrendered to the Allies. The humiliating Treaty of Versailles allowed Germany to keep six aging battleships, six cruisers, and some patrol boats, and banned the construction of new vessels. The few ships that were left, however, would not be needed much for travel abroad, for Germany also lost all its overseas colonies, and most other seagoing nations—especially those that had fought against Germany—closed their ports to German ships. For morale reasons, the German Admiralty was desperate to find a way to fly the German flag outside home waters, and in 1919 began planning a scientific expedition around the world.

The Germans had begun construction of a small gunboat in February 1914 in the naval shipyard in Danzig (now Gdansk, Poland). By the time it was launched one year later, though, World War I had erupted, and a small gunboat was all but useless. The Admiralty decided to convert it into a research vessel but could not afford to, because of demands to build more battleships, torpedo boats, and submarines. The little ship, named the *Meteor,* was mothballed at Danzig for the rest of the war. In the mad firestorm of rumors that followed the surrender, the Admiralty heard that the ship had been sold. The rumor was false, but the ship was quickly spirited from its berth in Danzig—which was to become a free city under the jurisdiction of the newly created League of Nations—and towed to Kiel, on Germany's Baltic coast. The German government then began negotiations with the Allies for permission to convert the *Meteor* to an oceanographic survey ship. After submitting plans that included expanded lab space, living quarters for scientists, and—most important—no guns, the Allies approved the conversion in October 1920. The ship was then towed to Wilhelmshaven, on the North Sea, where the refit began.

From 1921 to 1923 financial difficulties almost scuttled the project. Funds were raised to install diesel engines on *Meteor*, but the value of the money evaporated in the hyperinflation that crippled the German economy, and the engines were never installed. For a time all work on the ship was stopped until the value of the German mark ended its free fall at the end of 1923. Even though the Admiralty resumed the conversion of the ship afterward, the expedition itself lacked sufficient funding and appeared doomed.

The Notgemeinschaft der Deutschen Wissenschaft—the same organization that subsequently funded Alfred Wegener's final expedition to Greenland—began to discuss funding an oceanographic expedition in a meeting in February 1924. Even scientists had suffered from the humiliating defeat in World War I, and an expedition to foreign waters seemed to be the best way to restore the visibility and prestige of both the German navy and German science. Alfred Merz, director of the Institute of Oceanology in Berlin, was present at the discussion and immediately proposed an expedition to the Atlantic, focusing on the expanse between the tropic of Cancer south to the Antarctic ice pack. He was well prepared to seize the moment, for he had previously prepared a plan for a three-year voyage to the Pacific by the *Meteor* before the funding crisis led to its cancellation.

The plan was swiftly approved by both the research consortium and the Admiralty. The consortium would fund acquisition of scientific instruments, operating expenses for the voyage, and salaries and expenses of the expedition scientists, while the Admiralty would pay for the ship and crew.

Meteor, following an earlier four-week pilot voyage that led to several improvements to the ship and equipment, departed from Wilhelmshaven on April 16, 1925. The ship was equipped with deep-sea anchoring gear, protected and unprotected thermometers, water bottles, current meters, nets, analytical materials, two types of shallow-water echo sounders, and two types of deep-sea echo sounders, one of which was based on Fessenden's fathometer.

Captain Fritz Spiess, who had earlier been in charge of the Admiralty's Hydrographic Division, commanded the ship. Because of its small size, the ship's crew would have to be responsible for many of the scientific measurements and so was extensively trained: Some were sent to the Institute of Oceanology for advanced studies.

Merz directed the scientific activities. He was accompanied by his student and brother-in-law, Georg Wüst. (Ironically Fritz Loewe, who would later accompany Wegener on the 1929 and 1930 expeditions to Greenland, had been diligently involved in the planning for the *Meteor* expedition and had served on the pilot voyage but was dismissed prior to the main voyage. His assigned duties had included a determination of salt content in seawater using color as an indicator. Unfortunately, he was color-blind! In 1933 Loewe, a World War I veteran and Iron Cross recipient, spent a month in what the Nazis called "protective custody." He subsequently—and fortunately—went into exile, taking his wife and daughters to England. His crime: being a Jew.)

The expedition set out along thirteen evenly spaced profiles—or transects—across the Atlantic from about twenty degrees north latitude to about fifty-five degrees south, plus one more across the Gulf of Guinea. The echo sounders were constantly running, obtaining a sounding every twenty minutes. At intervals of 150 nautical miles, the ship was stopped for observations. Soundings were obtained by the traditional lowering of a weighted wire as a control for the depth data provided by the echo sounders. Temperature, salinity, and current measurements were taken from the sea surface to the seafloor. Water samples were taken for chemical analyses. Collections were made of marine organisms and ocean-floor sediments and rocks, and the characteristics of the atmosphere were also studied.

Not much sampling was planned for the initial leg of the voyage, from Wilhelmshaven to Buenos Aires, where sampling along the transects would begin. The *Meteor* reached Buenos Aires on May 24, took on coal, provisions, and supplies, and set out along the first profile, at about forty-two degrees south latitude. The ship wasn't out long, however, before it had to return to Buenos Aires. Merz had suffered a relapse of a lung ailment and had to be hospitalized. The *Meteor* arrived back in port on June 13. Merz was taken off. While there, another tragedy struck when a sailor returning to the ship fell off a bridge into the cold harbor water and died. After his funeral the *Meteor* returned to sea. In Merz's absence, Spiess took over official control of the scientific program, although Wüst ably handled most of the responsibilities. On August 25, as the *Meteor* was approaching the Brazilian coast toward the end of work on the second profile—at twenty-nine degrees south latitude—the crew learned that Merz had died. (Another sailor later died on the return trip to Germany, having apparently fallen overboard while sleeping on the deck.)

The weather in the "Roaring Forties" lived up to its reputation, but the sturdy little ship handled well. The fathometers proved their worth in an unexpected way. An exceptionally fierce storm battered the *Meteor* as it was in the vicinity of Gough Island. The officers were unable to determine the ship's exact position because of the storm—which increased the danger of the ship running aground—but increasingly shallow fathometer readings told them when *Meteor* was approaching land.

For the next two years *Meteor* zigzagged across the Atlantic, traveling 67,000 miles, worked 310 sampling stations, obtained 9,400 temperature measurements, and took more than 33,000 duplicated echo soundings—of which about 300 were confirmed by line soundings—in an area where only about 3,000 soundings had been obtained previously. A much more detailed and complex picture of the seafloor emerged from expedition data. A submarine ridge, part of the Scotia Arc, was discovered that cut off deep water flow between the South Atlantic and South Pacific Oceans. East of the Scotia Arc, a trench, named the South Sandwich Trench, was found. A large number of seamounts were discovered. The existence of deep currents in the ocean was confirmed.

The most important of *Meteor's* discoveries concerned the Mid-Atlantic Ridge—Matthew Fontaine Maury's old Dolphin Rise in the middle of the North Atlantic. Sounding data proved that the ridge continued into the South Atlantic between South America and Africa—except where sliced by a huge gash in the ocean floor, which the expedition also discovered and named the Romanche Deep. The ridge, the expedition found, further continued on around Africa into the Indian Ocean. It was also much more narrow than either Maury or Sir John Murray could have imagined with their sparse data.

The scientists and crew of the *Meteor* received a hero's welcome upon their return to Wilhelmshaven on June 2, 1927. The expedition helped to lift the veil covering the face of the deep, and it helped restore the prestige of the Fatherland and its scientists. Before those scientists had time to capitalize on the achievement, however, a madman would come along and tear it down.

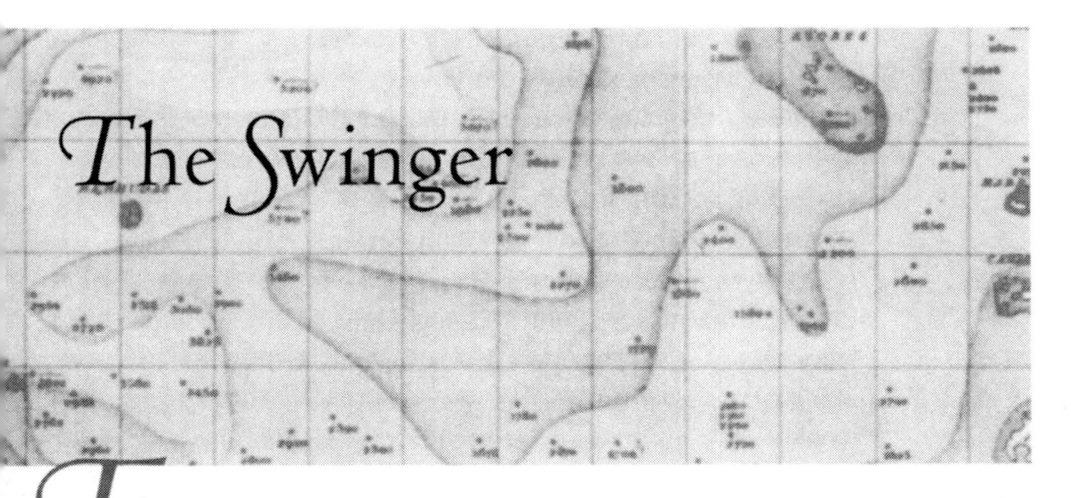

The Swinger

Felix Andries Vening Meinesz had a problem. The Netherlands State Committee for Leveling and Arc Measurements (later the Netherlands Geodetic Commission) had requested a gravity survey in 1912 as part of its effort to map the country. Vening Meinesz, who two years before had earned a degree in civil engineering from the Technical University in Delft, was commissioned to conduct it. He set up a number of stations where pendulum observations could be made, but vibrations of the unstable, waterlogged soil—compounded by the pounding waves of the stormy North Sea against the nation's shores—affected pendulum motion and made it difficult to get precise measurements. At least the measurements were not precise enough to make Vening Meinesz happy.

Any single pendulum apparatus, including the ones that Vening Meinesz initially used, is highly sensitive to motions of the surface on which it is mounted. Rather than make do with the equipment available, Vening Meinesz began designing something better. His solution was ingenious: By having two pendulums swing from the same axis but in opposite directions—as one swings to the left, the other swings to the right—the opposing movements would cancel the effect of surface motions.

His initial apparatus consisted of two pairs of pendulums set perpendicular to each other. A mechanism started one pair of pendulums swinging simultaneously while keeping the amplitude (the degree of swing from left to right) constant. The other pair would then be started in the same manner. It worked beautifully. For this invention Vening Meinesz received a doctorate from Delft Technical University in 1915.

Vening Meinesz realized the important contribution that his multiple-pendulum apparatus could make to the debate over the existence of isostasy. If multiple swinging pendulums could eliminate errors resulting from land movements, then maybe they could enable accurate and precise determinations of gravity at sea by eliminating errors caused by wave and ship motions. He further modified his apparatus and in 1923 was ready for a sea trial, taking it on a small steamer, the *Paleleh*. This time the experiment didn't go well. Wave motions at the surface were too great.

Even though his apparatus did not perform well on the ship, Vening Meinesz had other options. A colleague suggested that he study the feasibility of using submarines, which, when submerged, should not be affected by surface disturbances—even during rough weather. Vening Meinesz sought the assistance of the Royal Netherlands Navy, which agreed to provide submarines. The preliminary trials held in the vicinity of the Dutch naval base at Helder went well. It was time to test the equipment on a longer cruise.

For this voyage, Vening Meinesz added an additional modification to the apparatus. He previously had measured the period of the pendulum visually, but that would be impractical on a submarine. He added a photographic mechanism to record pendulum motions.

Vening Meinesz installed himself and his apparatus on board the submarine *K II*. On September 18, 1923, the *K II*, accompanied by two other boats, *K VII* and *K VIII*, and a support ship, the *Pelikaan*, departed from Helder on September 18, 1923, bound for the Netherlands East Indies (now Indonesia) by way of the Suez Canal. Submarines in the days before nuclear power had one big handicap: They could not run submerged indefinitely. While submerged the boats ran on battery power, but the batteries had to be charged at least once a day. When it was time to recharge, the boat would come to the surface and run on diesel power. In wartime, a sub on the surface could be easily detected and destroyed. In peacetime, the biggest problem a boat faced on the surface was weather. And during the *K II*'s voyage foul weather made the first few days—when most people's stomachs adjust to the motion of a ship at sea—quite unpleasant.

"Leaving Helder, we found at once a rough sea and the storm continued for five and a half days till we sighted Cape Villano on the N. W. coast of Spain on the evening of September 23," Vening Meinesz later wrote. "Only those who know submarine life, can imagine what life is like in these conditions. Even those who are sea-sick proof, suffer from the strong ship's movements, the stuffy air inside the ship, the wet clothes which refuse to dry in the humid atmosphere and numerous other small inconveniences."

The worst effect of the storm for Vening Meinesz, however, was that it kept him from making any gravity observations. Even when submerged, the disturbance caused by the large waves on the surface made the boat roll excessively.

The weather eased by the twenty-sixth, and that afternoon, with *K II* submerged, Vening Meinesz began testing his apparatus. For this he made two sets of measurements while cruising toward the Spanish coast, one while traveling east and one while traveling west. By doing so, he performed a powerful test of the sufficiency of his multiple-pendulum design. From measurements taken in both directions, Vening Meinesz would be able to obtain slightly different measurements of gravity due to the Eötvös effect—a change in the centrifugal acceleration of the earth's rotation due to the east-west component of the ship's speed. Traveling east with the direction of the earth's rotation would result in slightly higher centrifugal acceleration values, hence slightly lower gravity values. Traveling west, against the direction of the earth's rotation, would result in slightly higher estimates of gravity. At Gibraltar, Vening Meinesz worked up the results and found the effect. His apparatus appeared to be working—and working quite well.

Upon reaching an observation station, the submarine would submerge to the proper depth and establish a level course. A protective padding on the pendulums would be removed. The apparatus's light would be turned on, the position of the photographic and recording equipment adjusted. An electrical mechanism that controlled the camera shutter was adjusted so that the shutter would open at the proper time. Then the pendulums were lowered into position and their amplitudes recorded; one pair would be started, followed by the other. Once the photographic paper had been set in motion to record a series of exposures, the observations began and were run for fifteen to twenty minutes. At the beginning and end of the observation period, the temperature, atmospheric pressure, and relative humidity were measured.

Vening Meinesz was not the only one involved in taking observations. Steering a sub straight and level can be a challenge for the helmsmen; it is even more so if the crew is moving about the boat. Thus the entire crew had to remain still until the observation was complete.

After the storms on the first leg of the voyage, Vening Meinesz resolved to take advantage of the facilities at the British naval base on Gibraltar to devise a cradle in which the apparatus would be suspended. He hoped the new suspension system would eliminate the effect of rolling motions on the pendulums and permit gravity measurements during rough weather.

Departing Gibraltar on October 3, the submarines and their support ship sailed east across the Mediterranean, then passed through the Suez Canal into the Red Sea. While storms in the North Sea had made the crew miserable, the placid Red Sea had its own brand of torment. During that era ventilation was not included in a submarine's design—for good reason—and early subs had no air conditioning. Such characteristics, combined with a scorching tropical sun, can create a close, confined hell at sea. While submerged, the temperature in the boat rose to ninety-eight degrees with nearly 100 percent relative humidity.

STORMS AND SWELTERING temperatures weren't the only unpleasantness Vening Meinesz confronted. Submarines are not known for their luxurious accommodations. Space is always at a premium. Conditions were even more confined on the boats in the years between the two Great Wars. The tight fit posed a special challenge for Vening Meinesz. At a robust six feet six inches tall, no navy in the world would have allowed him to join the submarine service. The rules were more flexible for a scientist, but some were skeptical about the huge man's ability to cope in such tight quarters. When Vening Meinesz had showed up at the Dutch naval base in Helder for the preliminary tests in the summer of 1923, the commander of the first submarine he boarded said, "This won't survive a trial trip of twenty-four hours."

Vening Meinesz spent his days aboard doubled up "like a half-shut knife." His head had an unfortunate habit of getting in the way of bulkheads and equipment. There was no rest for the weary man, even while attempting to sleep. Harry Hammond Hess, who in 1932 would accompany Vening Meinesz on a cruise of the United States Navy submarine *S-48*, later wrote, "He slept in a bunk a foot shorter than he was, with only 18 inches to the bunk above. To turn over he had to climb out and get back in again." On some boats holes for his feet were cut in the bulkheads so he could sleep in a relatively straight position.

Conditions aboard a submarine also made it difficult to think well enough to evaluate the data adequately. Vening Meinesz and Frederick Eugene Wright of the Carnegie Institution in Washington, D.C., in their report on the results of a cruise of the United States Navy's *S-21* to the West Indies and Gulf of Mexico, wrote, "In spite of the novelty and interest of life aboard a submarine, it soon became evident to the scientists that the crowded conditions aboard, the lack of adequate ventilation, and poor air resulting in part from the Diesel engine gases coupled with tropical temperatures, are not conducive to intensive intellectual work and that computations and the working up of results are not possible at sea."

Vening Meinesz and Wright offered a piece of advice for those contemplating submarine work—leave the coats and ties at home. "For the benefit of future observers who may make gravity measurements aboard a submarine," they wrote: "It may be stated that khaki clothes are suitable for wear during the cruise. There is so much machinery and oil about that it is not possible to keep wearing apparel clean. Khaki is easily washed and serves the purpose well. An adequate supply of underwear should also be taken to provide for frequent changes in the tropical humid zone."

AFTER *K II* and its companions left the relatively sheltered waters of the Gulf of Aden, they encountered the large swells that roll north from Antarctic waters as well as more storms, but the pendulum's cradle performed admirably—even during rough weather, as Vening Meinesz had hoped. Vening Meinesz made an observation at his thirty-second and last ocean station on December 19 in the Strait of Malacca north of Sumatra. Afterward he transferred to the *Pelikaan* and continued on to Batavia (now Jakarta, Indonesia) on Christmas Eve.

For someone who wasn't supposed to last long on board a submarine, Vening Meinesz had done well. He had cruised halfway around the world, enduring weeks at sea in cramped quarters (especially for him), surviving repeated insults to his cranium, a sometimes violently swaying and pitching ship, and the shimmering but stifling climate of the tropics. The most important success, however, was that his apparatus had worked quite well, with a precision many times greater than had been achieved earlier in the century with hypsometers.

It was time to begin learning what gravity observations at sea would reveal about the nature of the earth. Vening Meinesz had obtained enough data from the *K II* cruise to contribute to one important controversy in geology: the nature of isostasy. At the time, two models of isostasy—both of which were developed decades before Clarence Dutton had actually coined the term—competed for supremacy.

WHILE CONDUCTING a survey of India, British Colonel George Everest—yes, that Everest—was puzzled by discrepancies in the calculation of the distance between two of his survey stations, Kalianpur in central India and Kaliana in the Himalayas. When measured by astronomical methods, the latitudinal difference between the two was about five seconds—or five hundred feet—less than when measured by triangulation.

In triangulation, horizontal and vertical angles between the surveying point and other objects were shot with an instrument, called a theodolite, which was mounted on a tripod for stability. Astronomical observations, made with instruments such as a sextant or octant, used vertical angles between the measuring point and objects in the sky to determine the location of a position on the ground. Regardless of how the angle data were obtained, the instruments had to be oriented properly if the observations were to be accurate. To achieve the proper orientation, a weight on a string, called a plumb line, was used to determine the vertical plane. If the plumb line deviated in any way from vertical, the instruments would not be properly oriented, and the angle measurements would be off—more so for the astronomical method than the triangulation method.

Pierre Bouguer, a French hydrographer and mathematician, had discovered the gravitational pull of mountains while on an expedition to Peru in the 1730s. This attraction could affect the behavior of a plumb line. In a completely flat area, the line bob should hang straight down, but, when close to a massive body such as a mountain, the line is deflected toward the body. The amount of deviation from vertical indicates the strength of the attraction.

Everest suspected that the discrepancy in distance measurements was due to problems with the plumb line and asked John

Henry Pratt—a Cambridge-educated mathematician and archdeacon of Calcutta—to examine the matter.

Pratt calculated the expected deviation of a plumb line from vertical in the Himalayas. The actual deviation turned out to be less than expected. In an 1855 paper in the *Philosophical Transactions of the Royal Society of London,* "On the attraction of the Himalaya Mountains, and of the elevated regions beyond them, upon the plumb-line in India," Pratt discussed the cause of the difference and concluded—as had Bouguer nearly one hundred years before—that the mountains were less dense than typical continental crust.

Sir George Biddell Airy, Great Britain's astronomer royal (and inventor of an eyeglass lens that corrects for astigmatism), pondered the "missing mass" reported by Pratt. In a review that immediately followed Pratt's 1855 paper, he suggested that the earth's crust floats on a more dense, but fluid-like, layer. Mountains, in Airy's view, were like icebergs with roots below the surface. He then proposed that the excess mass above is balanced by the mass displaced in the deeper layer by the mountains' roots. The level at which the balance is achieved is the "depth of compensation," which varies according to the thickness of the crust.

Pratt began studying additional data. He was surprised to find that along the Indian coastline the deflection of the plumb line was greater than expected, and he concluded that ocean crust must be more dense than that of the continents. Pratt offered an alternative to Airy's model. He suggested that the crust is of uniform thickness and the depth of compensation is uniform. In his model, differences in the density of the material that makes up the crust lead to the differences in elevation of features on the surface. He ascribed the density differences to differential cooling and contraction of the earth.

The Reverend Osmond Fisher, in his 1881 book, *Physics of the Earth's Crust,* fused the Airy and Pratt models into a comprehensive theory. While he recognized that oceanic crust was more dense than continental crust (as in Pratt's model), Fisher believed that Airy's model applied in most other cases. Like Airy, Fisher invoked the iceberg analogy. The earth's surface "is analogous to the case of a broken-up area of ice, refrozen and floating upon the water," he wrote. "The thickened parts which stand higher above the general surface also project deeper into the liquid below."

By the time Vening Meinesz began the *K II* cruise, Airy's model had been well established by researchers on land. By the time Vening Meinesz finished, he had the data to prove that the model applied to the ocean floors, too.

VENING MEINESZ redesigned his pendulum apparatus after the *K II* cruise. The new design featured three pendulums instead of four. One pendulum would remain motionless while the other two were set swinging. The modification allowed Vening Meinesz to calculate a correction for an effect caused by horizontal motion of the ship and made the task of computing the gravity values easier, quicker, and more accurate. The motion of each pendulum was recorded separately by a redesigned photographic mechanism. Vening Meinesz also improved the timekeeping capabilities and the suspension system.

The new pendulum apparatus was tested on a month-long cruise of the submarine *K XI* from Helder to Alexandria, Egypt, in October and November of 1925. It performed less accurately than expected because of friction between the pendulums and their supports but still worked fairly well.

Upon his return to the Netherlands, Vening Meinesz solved the friction difficulty and also set about to modify the suspension system. The previous suspension system had removed the effect of rolling motions, but controlling pitch continued to be a problem, and crewmen still had to remain immobile during an observation. Vening Meinesz mounted both the pendulum mechanism and the recording apparatus on a gimbal suspension that would keep the apparatus vertical despite the pitch and roll of the boat. With the new gimbal suspension the crew would be able to move about the boat, and the helmsmen would not have to concentrate so hard on keeping the submarine straight and level while observations were taken.

Vening Meinesz's third submarine cruise was more ambitious than the first two. He was to take the *K XIII* on a seven-month voyage from Helder to Java by way of the Panama Canal. Along the way he would have the opportunity to make observations over the Mid-Atlantic Ridge, the Puerto Rico Trench, the west coast of North America from Panama to San Francisco, several trenches (Renard Trench near Hawaii, Nero Trench near Guam, Yap Trench, and Philippine Trench), and the eastern part of the Netherlands East Indies. Gravity observations would be made at least once a day, and more frequently as the ocean-floor topography warranted. Deciding when to obtain more detailed observations was made easier by the use of a new echo sounder.

K XIII departed Helder on May 13, 1926. The first leg, to Horta in the Azores, was rough on all the men as intense storms knocked the boat about. Storms rarely bothered them for the rest of the voyage. However, Pacific swells and long periods of isolation wore on everyone.

"With the exception of the first trip, of a few days on the West coast of Mexico and near Guam no rough seas have been met," Vening Meinesz wrote. "Although with these few exceptions we did not have rough passages, the sea was generally not quiet enough for living on the lower deck of the submarine or for opening the hatches in this deck. This makes it impossible to take exercise and the air inside the sub gets stuffy and hot. Thus the longer crossings were a heavy strain, especially those between Las Palmas and Curaçao of 17 days and between Honolulu and Guam of 19 days. As we did not generally follow shipping routes, we seldom saw other ships: between Honolulu and Guam, for instance, no ship was met; the ocean appeared completely deserted and our small submarine world was isolated amidst blue skies and blue seas."

Not all was drudgery, however.

"The day of October 9 was a short one for us: At 9 o'clock in the morning we passed the 180th meridian and we jumped on October 10," Vening Meinesz wrote.

On November 24 the submarine crossed the equator for the first time. "Although the ship did not give much room for celebrating this event, we made the most of it thanks to the sea leaving the lower deck accessible. One of the officers and five of the men who had never been across it before, underwent the usual ceremonies."

The most memorable day of the cruise, however, was December 23, when *K XIII* arrived in Surabaya. The 23,000-mile cruise—at the time the longest ever undertaken by a submarine—was over.

Vening Meinesz was not going home just yet, however. During the first two weeks of February 1927, he began a gravity survey of the Java Trench, the characteristics of which were well known because of an earlier bathymetric survey by Dutch submarines using echo sounders. The cruise was short but eventful. After one observation *K XIII* dropped to about 260 feet, below the submarine's test depth—the point below which the boat was not supposed to be operated for fear of being crushed by the water pressure. The crew managed to recover and bring the ship up without damage.

(Probably the most frightening incident Vening Meinesz experienced on board a submarine occurred on the United States Navy's *S-48*, of World War I vintage, in 1932. During a dive off Jamaica, air vents on the boat's forward ballast tanks jammed open. The flooding took the boat 150 feet below its test depth. The commanding officer's quick actions saved the submarine. Just nights before, on February 3, Vening Meinesz had survived an earthquake in Santiago, Cuba, that damaged 80 percent of the city's buildings.)

In the spring of 1928 Vening Meinesz modified the device that spooled the photographic paper. Because of jams that had shredded the paper on the *K XIII*'s Pacific crossing, he had lost all observations from two of his stations. The experience was not one he wanted to repeat.

From June 1929 to February 1930 Vening Meinesz accompanied *K XIII* on three cruises through the Netherlands East Indies. Rough seas marred the first cruise, from June 12 to August 12. A radioman was washed overboard and saved by a life buoy that fortunately landed near him. The second cruise, from October 8 to November 14, circumnavigated the Celebes (now Sulawesi), and the third cruise, from January 2 to 14, 1930, looped around Sumatra.

VENING MEINESZ was not a particularly young man when he began his submarine voyages at age thirty-six. He was fifty-one when he took his final cruise in 1939. He had traveled 125,000 miles and made 844 gravity observations at sea at a time when only about 1,200 existed. He had discovered that isostasy prevailed over the oceans, that the Mississippi River Delta is in isostatic equilibrium, and that a band of negative gravity anomalies parallels the submerged margins of the continental shelves. All were significant, but none was as important as a discovery he made on July 5, 1926, on the *K XIII* cruise from Helder to Surabaya.

As the *K XIII* crossed the Puerto Rico Trench, Vening Meinesz found a strongly negative gravity anomaly toward the inside—the side nearest the neighboring island arc—of the trench. He discovered a similar anomaly in the Java Trench of the Netherlands East Indies on the later cruises of the *K XIII* in 1926 and 1927. These anomalies were huge, ten times greater in magnitude than what had typically been found on the surface of the earth. (He also observed that earthquakes seemed concentrated in the area of the negative anomalies and that volcanic activity, while not "in" the anomaly, closely parallels it.) It was difficult to reconcile this finding of "missing mass" in an area where the crust seemed to be bending downward rather than up, as in a mountain range. He doubted that the problem was with the theory of isostasy, however.

Vening Meinesz believed that the crust of the earth was elastic, and that it rested on a plastic layer below. Building upon a proposal by Paulus Pieter Bijlaard, a Dutch engineer, Vening Meinesz proposed that the crust, when highly compressed laterally, will at first thicken, then give, buckling downward in a tight band. The depression would create the negative gravity anomaly. A weaker band of positive anomalies would result from the rim of the depression where the crust flexes downward. He later suggested the downward buckles originated along descending arms of convection currents in the mantle. Philip Henry Kuenen, another Dutch scientist who devised an ingenious way to test the hypothesis in the laboratory, called Vening Meinesz's downward buckle a "tectogene."

The hypothesis explained many phenomena seen along island arcs and their associated deep-sea trenches. But it would be twenty years before anyone would be able to find out if the hypothesis was true.

THE ERUPTION of World War II ended Vening Meinesz's submarine expeditions. After the war others, most notably John Lamar Worzel of the fledgling Lamont Geological Observatory, undertook a number of gravity expeditions in submarines. Worzel and his students made 2,747 gravity observations during thirty cruises from 1947 to 1959. They expanded greatly upon Vening Meinesz's work, making the first detailed gravity surveys along the edges of continents and bolstering previous conclusions about the differences between continental and oceanic crust. Worzel's group also surveyed the Hawaiian Islands, a number of seamounts, and submarine trenches and midocean ridges. Their gravity research shed light on the shape of the earth.

By 1955 Anton Graf and Lucien LaCoste had independently developed spring-type gravimeters that were less fragile and easier to install and run than the pendulum apparatus. They also simplified the task of calculating the gravity results. Most important, the spring gravimeters, when mounted on a gyro-stabilized platform such as one developed by Worzel, could be used on surface ships. Worzel tested the Graf instrument on a surface ship at sea in 1957 during the International Geophysical Year. LaCoste's device was tested in 1958.

The new instruments eventually rendered obsolete the type of voyage that had made Vening Meinesz famous, yet they enabled researchers to fill in only a few blanks regarding what was already known about gravity variations at sea. Vening Meinesz and those who followed him beneath the waves had already sketched the big picture.

VENING MEINESZ was trapped in the Netherlands by the Nazis' brutal invasion of his country. Nazi forces occupied his house, turning it into officers' quarters, and forced him to live in his basement. While the "ruthless invaders"—as he called them—enjoyed themselves upstairs, the steel-nerved Vening Meinesz held meetings of the Dutch resistance downstairs.

Despite the tribulations of the war, Vening Meinesz's spirit, and his sense of humor, never diminished. Nor did his interest in science. After the war he visited Maurice Ewing and his wife, Margaret, in their home on the grounds of Lamont Geological Observatory. The two men could not stop discussing geophysics, and Vening Meinesz repeatedly apologized to Mrs. Ewing for it.

When she said she was accustomed to shoptalk, Vening Meinesz replied, "But we are not talking shop, we are talking a whole department store."

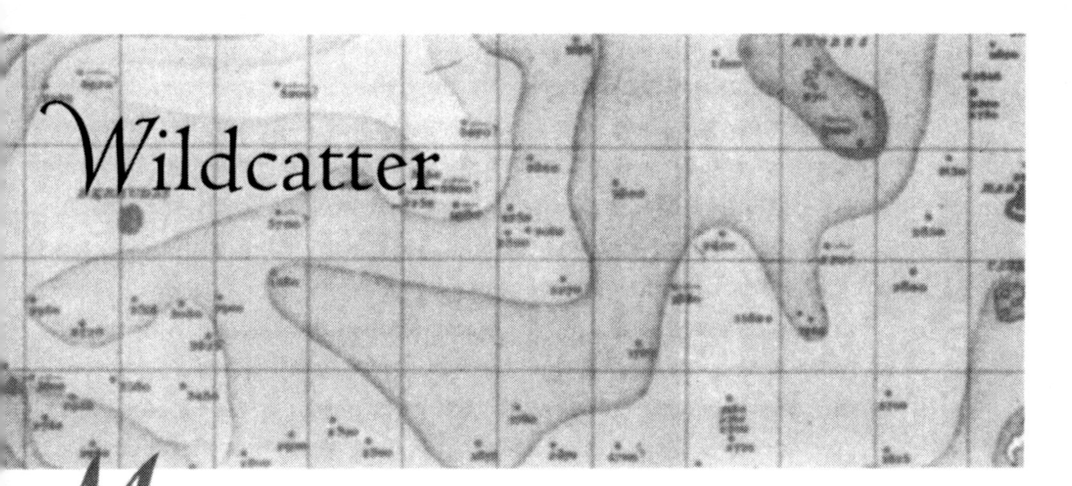

Wildcatter

Maurice Ewing grew up on an ocean of sorts—a harsh, unforgiving expanse of grass on the Llano Estacado, or Staked Plains, in the Panhandle of Texas. The Llano Estacado is a place of extremes. The clouds bless the farmers with adequate rainfall for a few years, then withdraw their favor with withering droughts and towering dust storms that seem to have no end. Broiling heat in the summers alternates with deadly Arctic-spawned blizzards in winter. Despite the fickle landscape the settlers came, built houses, farms, and ranches, and figured out how to survive. Maurice Ewing's parents, Floyd Ford Ewing and Hope Hamilton Ewing, did their best.

When Floyd Ewing, a cowhand at the F Ranch, became engaged to Hope Hamilton in 1899, the young cowboy knew that he would have to give up his forty-dollar-a-month job and become financially independent. In 1900, just a few months before the wedding, he left the ranch. Early in the spring of 1901 he went to New Mexico with a friend, Philip Fryer, to scout out land to claim as a homestead. Their initial impressions were not favorable. Water was scarce. The two men also had reason to suspect that cattle rustling was a popular pastime. They quickly returned to Texas.

After his wedding, in April 1901, Floyd Ewing found a small job tending a pump used to water another rancher's cattle, but the work soon ended. With no job, no land, and the costs of leasing range eroding what little profit they made with the few head of cattle they had, Floyd and Hope Ewing in 1902 made the difficult decision to leave Lockney, Texas—where they both had grown up—for a homestead eight miles away from the town of Elida in Roosevelt County, New Mexico.

The timing of the move proved disastrous. The area was in the middle of a five-year drought cycle. Crops failed. Cattle, desperate to find forage on the parched range, began to eat locoweed and died. In December 1903 Hope Ewing, who was pregnant with twins, fell. The accident triggered an early labor, and the babies died before a doctor based in Elida could reach the Ewings' home. The following June, Floyd Ewing was thrown from his horse after it stepped in a prairie dog hole. He broke his shoulder in the fall. In July, while he was recovering from that injury, their two-year-old son, Jack, drowned while playing in a watering tank for the cattle.

Early that fall the Ewings abandoned their half dug-out, sod-roofed home and returned to Lockney.

Maurice Ewing, whose rarely used first name was William, was born on May 12, 1906. Six brothers and sisters followed over the next seventeen years. The Ewing children grew up in a stimulating household. The father was an accomplished fiddler with a knack for storytelling. The mother was small, lively, and idealistic. The two adults prized education. The importance of learning was not lost on the children. Six of the seven who lived to adulthood went to college. The seventh, a daughter who married young, became a piano teacher. The record was impressive for a family whose head struggled to make a living farming and selling hardware and farm equipment.

EWING'S GRADES in grammar and high school indicated little aptitude for science and math, but an influential high school math teacher turned him "from a person who could do no math problems to one whose favorite thing was algebra," as Ewing recalled late in his life. When Ewing was ready to go to college most schools turned him down, telling him to study hard for another year and try again. A spirited recommendation from his math teacher, however, convinced Rice Institute (now Rice University) in Houston to offer the sixteen-year-old a scholarship.

Houston is a long way from the Panhandle, so Ewing had to travel as economically as possible. One year he spent twelve dollars to buy a motorcycle from a man who had taken it apart, but could not reassemble it. Ewing put it back together, then left for school with ten dollars in his pocket and a bedroll tied to the seat behind him. Before long the chain broke; then the motorcycle ran out of gas. Ewing abandoned the bike, hopped a freight train, and fell in with two hoboes. The trio was discovered by a brakeman, who intended to rob Ewing of his money and watch. Ewing, with a persuasive—but true—story about being a poor college student, talked the man into returning the cash and timepiece.

The ordeal was not over, however. Eventually the hoboes attacked Ewing. Despite being hit by a blackjack, he feigned insanity long enough to get away—but not without a chase. The attackers lost track of him as he hid in a churchyard. Most of his clothes and bedroll were gone, but he still had his ten dollars and the watch. Embarrassed about the state of the few clothes he had left upon his arrival in Houston, he would not board a streetcar. He instead talked the police into driving him to campus.

Ewing worked his way through school. His first job was at an all-night drugstore, his main responsibility being to take coffee and sandwiches to prostitutes living in the hotels around the Humble Building. He subsequently obtained part-time jobs with the university, grading papers and working in the library. During the summers, however, he had to find other ways to earn money. Initially he worked in a grain elevator. Later he began working with oil exploration crews in Louisiana. There he was exposed to gravity surveys and seismic profiling that paid big dividends in locating salt domes—where large reserves of oil might be found. Ewing's experience in the oil fields was to later pay even greater dividends for earth science.

EWING INITIALLY majored in electrical engineering at Rice; he doubted the wisdom of his choice.

"The engineering students I had noticed used a funny paper with red lines down and across it. They were careful how they made their figure fives. An inkblot was a catastrophe," Ewing later said. "For half a year, I studied the upperclassmen. The engineers were all worried."

Physics seemed more interesting—many of the classic papers of quantum mechanics appeared at the time—and the teachers were more approachable. Ewing, despite reservations about how he might make a living in physics, made the switch, not only in the degree he sought but also in frame of mind. He acquired a preference for simple arguments; detailed, well-understood, and carefully explained theories; and ingenious, frequently self-designed instruments.

Ewing's powers of observation were displayed one night on the Rice campus as he and another student walked back to their rooms after a long study session in the library. Glimpsing a multicolored, circular band of colors in the dew gathering on the grass, Ewing asked the other student to shuffle his feet in the grass, directing him along the path of the rainbow-like band. Ewing made notes of the pattern of the path and the position of the moon and soon produced his first published scientific paper, "Dewbows by Moonlight," which appeared in the journal *Science* in 1926.

After graduating in 1926, Ewing heeded what was probably the worst advice he ever received: One of his beloved mentors in the physics department told Ewing to stick to theoretical physics because he did not have any aptitude for empirical studies. The next five years of graduate study at Rice were productive for Ewing, however, and strengthened the foundation of his later life as . . . an empiricist! His dissertation, inspired by his summers exploring for oil, was titled "Calculation of Ray Paths from Seismic Travel-Time Curves." With the thesis Ewing helped place seismic refraction studies—in which the behavior of sound waves as they travel from a source near the surface to a distant listening point are used to reveal details of the earth's structure—on solid mathematical footing. (Seismic reflection, in which the sound source and listening point are close to each other, is analogous to echo sounding in water. Sound waves are partially reflected back from interfaces between rock layers and reveal more detail about strata beneath the listening point.) Ewing earned his Ph.D. from Rice in 1931.

In 1929 Ewing accepted a position as a physics instructor at the University of Pittsburgh. One year later he moved to a similar position at Lehigh University in Bethlehem, Pennsylvania. Despite the heavy teaching load, Ewing began a series of projects in geophysics—his primary interest being in the development of seismic research techniques using explosives. Most of his work was fairly minor in scope, but that changed on a cold, snowy day in November 1934 when two men in derby hats and long coats with fur collars visited Ewing at his lab in Lehigh.

The men, Richard Montgomery Field, of Princeton University, and William Bowie, who had been a participant in the 1926 symposium on continental drift in New York City, were both deeply interested in marine geology. Field, a man with the vision, and influence, to set the agenda for American research on the geophysics of ocean basins for decades, was chairman of the American Geophysical Union (AGU) committee created to study the matter. Bowie, also heavily involved in AGU affairs, was in charge of the United States Coast and Geodetic Survey, an agency charged with understanding the characteristics of the continental shelves. The pair made Ewing an offer he could not refuse.

"They said they wanted to interest me in the study of the continental shelf," Ewing said. "They thought it was a very important geological problem to see if the steep place where the shelf ends was a geologic fault or the result of outbuilding of sediment from the land—was it a basic geologic feature or a superficial appearance? And they wondered if seismic-refraction measurements, such as I had been working with, could be used. I said yes, it could be done, if one had the equipment and ships. If they had asked me to put seismographs on the moon instead of the ocean floor I'd have agreed, I was so desperate for a chance to do research. Bowie and Field thought the Geological Society of America would back me—'especially,' they said, 'if you don't mention us.'"

Ewing—who in fact did arrange to put seismographs on the moon during the Apollo missions of the 1960s and 1970s—followed the advice of Field and Bowie. It paid off in a two-thousand-dollar grant from the GSA. At the time, explosion seismology—a technique used to measure the thickness of the Greenland ice cap during Alfred Wegener's final expedition to the island—had not been attempted at sea. The plan was to shoot a number of seismic lines on the land from Richmond, Virginia, east to Cape Henry and out across the Atlantic to the edge of the continental shelf, about seventy-five miles out. There was little time to prepare. Ewing, assisted by his graduate student, Albert Paddock Crary, and H. M. Rutherford, who ran a seismological station in Pittsburgh, worked all night the last day of the spring semester getting ready for the expedition.

The next morning Ewing drove off alone in an overloaded panel truck, with bad tires, cast off by a local power company. Crary and Rutherford set off in Crary's car to Wilmington, Delaware, to pick up two hundred pounds of TNT from DuPont. After repeatedly being denied access to bridges—frequently encountered in a region repeatedly cut by streams as well as the mighty Chesapeake Bay—the pair tore from the car signs thoughtfully placed by DuPont employees warning of "High Explosives" and, after a long delay, began to progress south. However, the lack of sleep caught up with them between Richmond and Norfolk, Virginia, about five the next morning when Crary dozed at the wheel and rolled his car in a ditch.

Crary and Rutherford were uninjured but needed help. Looking back, they saw headlights approaching and, in the early morning twilight, realized that the lights were coming from Ewing's truck. But Ewing, hunched over with a verge-of-unconsciousness death grip on the steering wheel, did not seem to notice their attempts to flag him down. Five minutes later, however, he returned, feeling his presence was required up the road a piece, although he wasn't sure why.

All the men, vehicles, and equipment eventually made it to Norfolk, but Ewing was promptly taken to Virginia Beach to meet with the captain of the Coast and Geodetic Survey ship *Oceanographer* (the former yacht *Corsair*, which had been donated to the survey by John Pierpont Morgan). Ewing kept blacking out throughout the meeting. At least he got back to Norfolk in good shape. The captain was severely injured and an assistant—who would have assisted the Ewing group—killed in an automobile accident on their way back.

The loss of the captain was a crucial blow to Ewing's effort. "The written orders to the ship's captain were to allow us on board and—this for the eyes of congressional busybodies—do experiments for us on a not-to-interfere-with-ship's-work basis," Ewing said. "Explosives were used occasionally in the course of the ship's work, but it was understood that the orders were to be interpreted liberally."

The executive officer, who took over in the captain's absence, was not the liberal type. As far as he was concerned, Ewing's experiments were not to interfere with *Oceanographer*'s work, period. Ewing managed to obtain only a few seismic shots at night, while the ship was anchored. Scientifically, the results of the trip were disappointing, but the experience convinced him that seismic studies of the ocean floor were possible.

When *Oceanographer* returned to Norfolk, Ewing took a train to Wilmington, Delaware, to discuss the results with Field. Field suggested that Ewing continue shooting the land portion of the transect while Field discussed the possibility of obtaining ship time from Henry Bryant Bigelow, director of Woods Hole Oceanographic Institution (WHOI).

"Persuading them to carry explosives on their ships for the first time can't have been an easy job, but when we finished the land part we met him up there," Ewing said. "We were regarded with some suspicion, but were told that we could have a two-week cruise on *Atlantis*—a 146-foot steel sailing vessel that had been especially built for Woods Hole—after the regular season was completed."

Woods Hole officials apparently had scant confidence in Ewing, Crary, and Rutherford's ability to function on the open ocean. Columbus O'Donnell Iselin II was sent along to manage *Atlantis*'s affairs on the October 1935 cruise "in case we became useless," as Ewing recalled.

"Seasickness is like toothache, you know—you don't notice it if your house is burning," Ewing said. "This was the chance of my life, as far as I was concerned. We sent Crary off in the whaleboat with a couple of sailors to make the shots. He took a pile of blasting gelatin and dropped the biggest charge eight miles out. The others went at regular intervals on the way back to *Atlantis*, where he had geophones on the bottom. Crary had a radio and signaled us at each shot. Columbus seemed a little surprised when the messages kept coming back clear, but when the whaleboat was close he looked and said 'He's seasick now.' I had to tell Columbus that Crary was just chewing tobacco.

"We had our brush with seasickness later. Rutherford was down below developing one of the records, with both arms in the sleeves of a portable darkroom; he suddenly turned green and said he was going to be sick, what should he do? I drew back my arm and said, 'I don't know, but if you ruin that record I'll break your jaw,' and he didn't have any more trouble."

Ewing's seismic profiles revealed that the basement rocks that outcrop along the surface near Richmond slope gently downward toward the edge of the continental shelf, where they lie buried beneath a blanket of sediment twelve thousand feet deep. The finding was important—it was the first time anyone had presented any actual data on geological characteristics of strata beneath the ocean floor.

BY THE TIME the results of the seismic cruises were published in 1937, Doc, as Ewing was called by students and colleagues alike, had acquired two new students with whom he would be associated for years—in one case for the rest of his life. Allyn Collins Vine began graduate studies in physics at Lehigh in 1936. The same year John Lamar Worzel—who, as a result of a popular song of the day, acquired the nickname "Joe"—began studying mechanical engineering as an undergraduate at Lehigh.

Vine and Worzel knew each other from the boardinghouse where they lived. Both were wizards at designing instruments. Vine, as a youth in Garrettsville, Ohio, led raids on telephone company junkyards in search of wires and components to build devices like burglar alarms. Worzel, an iron man who survived jaundice, surgery for a double hernia, rheumatic fever, a tonsillectomy, mumps, measles, and peritonitis from a burst appendix—all by the time he turned ten—loved to take things apart to see how they worked.

Worzel was also interested in photography, and he was drawn into Doc's orbit in October 1936 when Vine offered to purchase Worzel's Argus camera. Vine wanted the camera to make some base gravity measurements following a gravity expedition to the Puerto Rico Trench by Ewing and Harry Hammond Hess aboard the navy submarine *Barracuda.* Worzel sold the Argus for the cost of a replacement. In the spring he began helping Vine and Norman Webster, another of Ewing's students, prepare illustrations for a paper on seismic measurements at sea. Soon after, Worzel changed his major from mechanical engineering to physics.

Late on a Friday night in October 1937, Vine walked into Worzel's room and asked if he would be interested in joining Doc, Vine, Webster, and George Prior Woollard, of Princeton University, on an excursion to shoot seismic profiles of the coastal plain of New Jersey. Worzel would have to cut a Saturday morning class if he went along—for the privilege of helping dig holes six inches wide and ten to forty feet deep in which to place the explosive charges—but, like a disciple asked to abandon everything for the chance of a life with the prophet, he agreed to go along.

AT FIVE the next morning Ewing and his entourage—Vine, Webster, and Worzel—piled into Doc's 1934 Ford sedan named Floozey Belle. The car had two front bucket seats, a removable back seat, two side doors, and a rear door. Floozey Belle ordinarily seated five when both seats were in place. On the way out of Bethlehem that morning the car, without the back seat, still managed four passengers—three divided between the two bucket seats in front and one squeezed in the back along with an oscillograph, amplifiers, geophones, and cables. Hand augers, extension pipes, and a reel of conductor wire were strapped to the outside.

Floozey Belle's initial destination was Princeton, where they were to meet Woollard and his Model A filled with cases of dynamite. From there they proceeded to the site of their first shot, just outside the town. At each location the men first laid out an array of geophones, then laid conductor wire to the shot point. At the shot point the men dug a hole as close as possible to the boundary line dividing private property from public right-of-way. That way, if a police officer interrupted them, they could argue they were on private land, whereas if a landowner accosted them, they could say they were on the right-of-way.

Traffic was another problem Ewing's team had to confront. Passing cars could be struck by debris from the blasting, and the noise and vibration from vehicles could interfere with the geophones. "One Saturday afternoon . . . Woollard stopped a car coming along the road, saying with his charming Southern accent, 'Sorry, sir, we're about to do some work down here. You'll have to wait a few minutes,'" Worzel wrote afterward. "In a rush, the driver protested, and Woollard reached in and took his keys. When we got our shot off, Woollard said, 'Thank you, sir' and returned the keys."

Ewing would be notified by telephone once the holes were dug and the charges placed. He would then warm up the amplifiers, wait for the geophones to quiet down, and give the order to shoot to whomever had the detonator. Doc hand-cranked photographic paper through the oscillograph until the noise from the blast had reached all the geophones. The photographic paper would be developed in the field by putting it into "Minnie's drawers"—essentially a darkroom in a sealed black bag with sleeves to insert one's arms in—and drawing it through a light-tight box containing cans of developer, water, and stop bath. A quick glance at the quickly developed paper told Ewing how far the group should travel before setting the next shot.

Throughout the fall of 1937 Ewing and company continued building a seismic profile of the coastal plain of New Jersey on a line from Princeton to Barnegat Bay. One mishap in Silverton came close to bringing their effort to ruin when a bread truck ran into the rear fender of Worzel's Ford coupe: At the time the rumble seat of Worzel's car was filled with dynamite. Fortunately, the damage to Worzel's car amounted to only twenty dollars—all of it confined to Worzel's fender. No one among the crowd of onlookers—including one very nosy body shop worker—discovered the explosives. The bread company paid for the repairs.

The schedule was grueling. Ewing and his assistants would work late into the night on Fridays getting the equipment ready and loaded. Then they would leave by five the next morning (or earlier, depending on how far they had to drive) in order to get to their first site by sunrise. The men would eat—but not stop for—lunch, working throughout the day until sunset. Finally, they would find a boardinghouse, eat dinner, then begin evaluating the day's result. The photographic paper containing the results would be thoroughly washed—often in a bathtub in the rooming house. The records would be read carefully, the data plotted on graph paper, and calculations made of the thickness of and sound velocity in the strata below the surface. Finally, plans for the next day would be made, and the men would turn in at midnight—and start again at five on Sunday morning. They would work again until dark, eat in a diner on the way back to Bethlehem, and unload the cars—it was especially important to get the gear out of Floozey Belle, as Ewing's family needed it during the week. The men would finally get to bed in the early hours of Monday morning, not long before they had to be in class.

EVEN WHILE setting off bombs about the New Jersey countryside, Ewing turned his attention to the structure of the ocean floor beyond the edge of the continental shelves. The challenges were numerous. Ewing's initial approach involved lowering an oscillograph, geophones, and explosive charges to the ocean floor. All the equipment—even the containers for the explosives—had to be designed by Doc and his students to withstand pressures of eight thousand pounds per square inch and temperatures approaching freezing.

In addition to designing their instruments, the group also had to design the equipment needed to test their handiwork. They devised a pressure chamber from a surplus fourteen-inch naval artillery shell—the sharp end was buried in a hole in a ground-floor laboratory in the physics department building at Lehigh. A one-ton hydraulic jack was joined, through a hole in the back of the shell, to a plug that would ordinarily have sealed the powder inside. A former student of Ewing's devised a gasket for the chamber, which, when filled with oil, could generate pressures as great as ten thousand pounds per square inch. Such ingenuity was characteristic of the Ewing group. When he and Vine were designing a deep-sea camera, they searched for a way to protect flashbulbs from the pressure at the ocean floor. They found their answer in a diner—rather, in a diner's glassware, which made superb transparent pressure vessels. Another time someone in the group discovered that fruit salad cans were the perfect size for protecting one of their instruments at sea. "So we all ate a lot of fruit salad," Worzel recalled, "and then we nickel-plated the cans so that they wouldn't rust in the ocean."

The deep-sea seismic equipment was first tested on two cruises of the *Atlantis,* in July and September of 1937. When it was time to attempt a shot, flaked TNT would be melted on the deck of the ship in a contraption called "Vine's still"—which used steam from the engine room as a heat source—and poured into a protective casing. An electronically driven detonator, attached to leads on the cover of the container, was included with the charge.

(Another protégé of Richard Field, British geophysicist Edward Crisp "Teddy" Bullard—later Sir Edward, or "Sir Teddy"—was also on one of the 1937 cruises. One night, while he was watching Ewing melt down the TNT, *Atlantis* Captain Frederick McMurray walked up and began knocking out his pipe on the side of a nearby box. "Ewing looked at him and, after what seemed like a long time, said, 'You know Captain McMurray, if I was you I wouldn't knock my pipe out on that there box,'" Bullard later recalled. "McMurray continued to knock out his pipe and said, 'And for why Dr. Ewing, if you was me, wouldn't you knock your pipe out on this box?' I could stand it no longer and said 'That box is full of TNT.' McMurray didn't say a word. He just turned round and walked out.")

The array of equipment—oscillograph, geophones, then three explosive charges, each about one thousand feet apart, at the end—was lowered by cable to the ocean floor. When the crew thought the charge at the end had neared the bottom, *Atlantis* would get underway so that the equipment would lie in a line rather than piled on top of itself. Once the equipment was deployed, the ship would stop and the bombs would be fired in sequence beginning with the one at the end of the line. Finally, the remainder of the array would be recovered. The complicated process usually turned into an all-day affair.

None of the four deep-sea seismic shots taken on the 1937 cruises was successful. Ewing, however, was not one to give up. In 1938 he began experimenting with free-floating bombs and equipment. The geophones, recorders, and firing clocks were attached to balloons filled with gasoline for buoyancy. The bombs, equipment, and some ballast would be dropped overboard, given time to sink to the bottom, and the bombs set off. A "timer" consisting of block salt would "count down" (dissolve) until ballast-release time. When the salt was fully dissolved, ballast would drop from the equipment so that the balloons could float it back to the surface. The new rig was tested in the waters off Bermuda during the winter and spring of 1938 and 1939. The float system worked, but now the oscillographs failed.

Testing continued the next year when Worzel accidentally discovered the cause of failure. He had been using Eveready batteries to power the oscillographs when testing them in the lab but then replacing the Evereadys with Burgess batteries when it was time to make a shot. The one time Worzel forgot to make the switch—on the final shot of the year in September 1940—the oscillograph worked.

During the last half of the 1930s Ewing and company had already made major contributions to marine geology and geophysics. Ewing and Vine had developed the first working deep-sea camera and had taken dozens of photos of the abyssal floor. Doc himself had improved the timekeeping apparatus for the Vening Meinesz pendulum apparatus. Ewing's team had obtained the first solid data on the geological characteristics of the continental shelf and, by the end of 1940, had solved many of the problems of obtaining seismic profiles from the deep-ocean floor.

In September 1939 another war erupted in Europe, and soon storm clouds loomed off the American shores. Geology would have to wait. More pressing matters required attention.

A Rumor of War—and the Real Thing

Both Columbus Iselin, who succeeded Bigelow as director of the Woods Hole Oceanographic Institution, and Maurice Ewing had a feeling that the United States would be drawn into the two conflicts raging in Europe and the Far East. After the *Atlantis* returned to Woods Hole in September 1940, Allyn Vine and Joe Worzel loaded up Floozey Belle and a dump truck with their equipment from the cruise and with some rusting machine shop tools purchased from the owner of a service station next to WHOI. The service station had been flooded during the Great Hurricane of 1938 and the equipment, rarely used before the storm, was completely neglected after. While the two slowly made their way back to Bethlehem, Ewing stayed behind for talks with Iselin.

Not long after their arrival in Bethlehem—after, of course, the shop tools (including a lathe, milling machine, and drill press) had been unloaded, cleaned, and reassembled in their lab at Lehigh—Vine and Worzel got a telephone call from Ewing. Doc was taking a leave of absence from Lehigh to participate in a military research program at Woods Hole to be funded by the National Defense Research Committee (NDRC). Vine and Worzel were welcome to come along if they wished.

The two students discussed Doc's offer. Vine had finished his Ph.D. and was preparing to take his orals. Worzel was about to begin graduate studies at Lehigh—with a scholarship to pay for his tuition. But there was only one thing they could do.

"We hired another dump truck, loaded up the machines, and set them up in a basement at WHOI," Worzel wrote. "Those three heavy machines had a ride of more than eight hundred miles to get across the fence at Woods Hole."

The machines formed the nucleus of the first machine shop at WHOI. The equipment proved crucial for the work that was soon to follow.

BY 1940 Fessenden's echo sounder had evolved into sonar (SOund Navigation And Ranging), a powerful tool for detecting hostile ships near one's own ship. Ewing and Iselin readily appreciated that sound traveling through water could produce confusing signals. Whether on a surface ship hunting submarines or on a submarine hunting surface ships, knowing the behavior of sound in water would be crucial to a crew's survival—or another's demise. They began assimilating all that was known at the time. Thanks to Ewing's seismic research, they were best qualified to undertake such an important study.

The timing could not have been better. Atlantic shipping was already suffering from U-boat attacks. (In fact, the first United States Navy vessel sunk during the war was not at Pearl Harbor but in the North Atlantic when the destroyer *Reuben James*—escorting a convoy about six hundred miles west of Ireland—was torpedoed by *U-552* on the morning of October 31, 1941.) On December 7, 1941, just hours after the Japanese surprise attack on Pearl Harbor, United States subs were ordered to execute unrestricted submarine warfare against Japan.

As soon as Worzel arrived back in Woods Hole, he and Ewing began studying the data on sound intensity versus distance that had been previously collected on *Atlantis*. Ewing soon demonstrated that various confusing characteristics observed could be explained by the refraction, or bending, of sound in seawater. Refraction is caused by variations in the velocity of sound under changing water conditions. Sound velocity, they knew, is primarily controlled by water temperature and, to a lesser extent, by salinity and pressure. Ewing and Iselin decided to write a report explaining the refraction of sound waves under a variety of conditions at sea. The report would include instructions on how to calculate refraction patterns from temperature, salinity, and pressure measurements and would summarize variation in sound behavior by season and among different marine environments. Since salinity typically had only a minor effect on sound refraction, Ewing and Iselin concluded that the determination of temperature versus depth was sufficient to assess local sound conditions.

Iselin began writing a description of the marine environment in regions where the war was then being fought and where it would likely be fought as events unfolded. Ewing began preparing a discussion of the physics of sound refraction at sea, and Worzel began drafting the graphs and tables needed to make the refraction calculations and also figures to explain the intensity versus distance observations already obtained.

Over the next few weeks Ewing and Worzel collaborated with Iselin and by December 1940 produced a definitive manual for the navy, *Sound Transmission in Sea Water.* One night Worzel was plotting the results of ten thousand calculations he had made to determine sound velocity under a full range of distance, depth, pressure, salinity, and temperature values.

"I started on the work one evening after dinner," Worzel wrote. "Ewing came in, saw how tedious the work was, and sat down to help. When we finally finished at six o'clock the next morning, Ewing looked at me and said, 'You know, Worzel, if you hadn't finished this by yourself by midnight, I would have given you hell.'"

Among the phenomena Ewing and company described was the divergence of sound waves from a source on the surface: Part of the signal bends back toward the surface while the other part bends toward the bottom. In the "shadow zone" between, a submarine could lurk undetected by sonar. They also discovered that sound waves tend to bounce off thermoclines—zones of sudden change in water temperature, generally a steep decline in temperature with depth. By descending below a thermocline, submarines could again find shelter from their hunters above. Ewing also discovered the sofar (SOund Fixing And Ranging) channel, a layer of water into which sound waves are repeatedly refracted so that the waves effectively disperse in two dimensions rather than three. A sound in the sofar channel can be heard sometimes thousands of miles away. (In one experiment in the 1960s a detonation of about two hundred pounds of TNT near Australia was heard in the sofar channel off Cape Hatteras, North Carolina.)

Now that the effects of temperature on sound transmission in water had been analyzed, an instrument was needed that could rapidly document the change in temperature with a change in depth. Carl-Gustav Rossby, a Swedish-born professor of meteorology at the Massachusetts Institute of Technology, in 1934 had designed and built a difficult device—called an oceanograph—that was supposed to continually measure the temperature of water while being dropped from the surface to a depth of several hundred feet. Rossby did not overestimate his design skills, however, and eventually gave the device to a friend, South African Athelstan Frederick Spilhaus, to come up with something better. Shortly thereafter Spilhaus did, applying in 1938 for a patent on an instrument he called the bathythermograph—BT for short. Spilhaus, by coupling the BT with a stylus that dragged across a slide of smoked glass, made it possible to record the temperature profile of the water column. Even his device, however, took too long to register temperature changes; thus it could be used only from a ship that wasn't moving.

While work progressed on *Sound Transmission in Sea Water,* Ewing and Vine began redesigning the bathythermograph. Their BT design could respond to temperature change more quickly, could measure water pressure (thus indicate depth), and could be lowered and raised quickly on a fast-moving ship, such as a destroyer on the hunt for a submarine. About one year later they designed a BT for

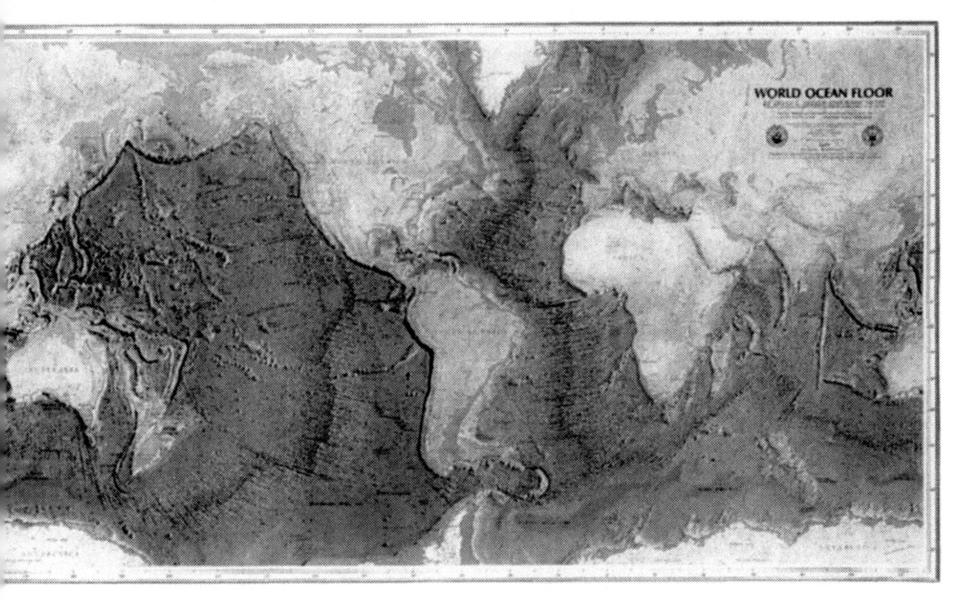

▲ The *World Ocean Floor* map by Bruce C. Heezen, Marie Tharp, and Heinrich Berann. Copyright 1977 by Marie Tharp. Reproduced by permission of Marie Tharp.

▼ Alfred Wegener in a thoughtful moment, Koch expedition to Greenland. Copyright Arctic Institute, Copenhagen.

The last photo of Alfred Wegener and Rasmus Willumsen, before they departed Eïsmitte, 1930 expedition to Greenland. Copyright Alfred Wegener Institute for Polar and Marine Research, Bremerhaven, Germany.

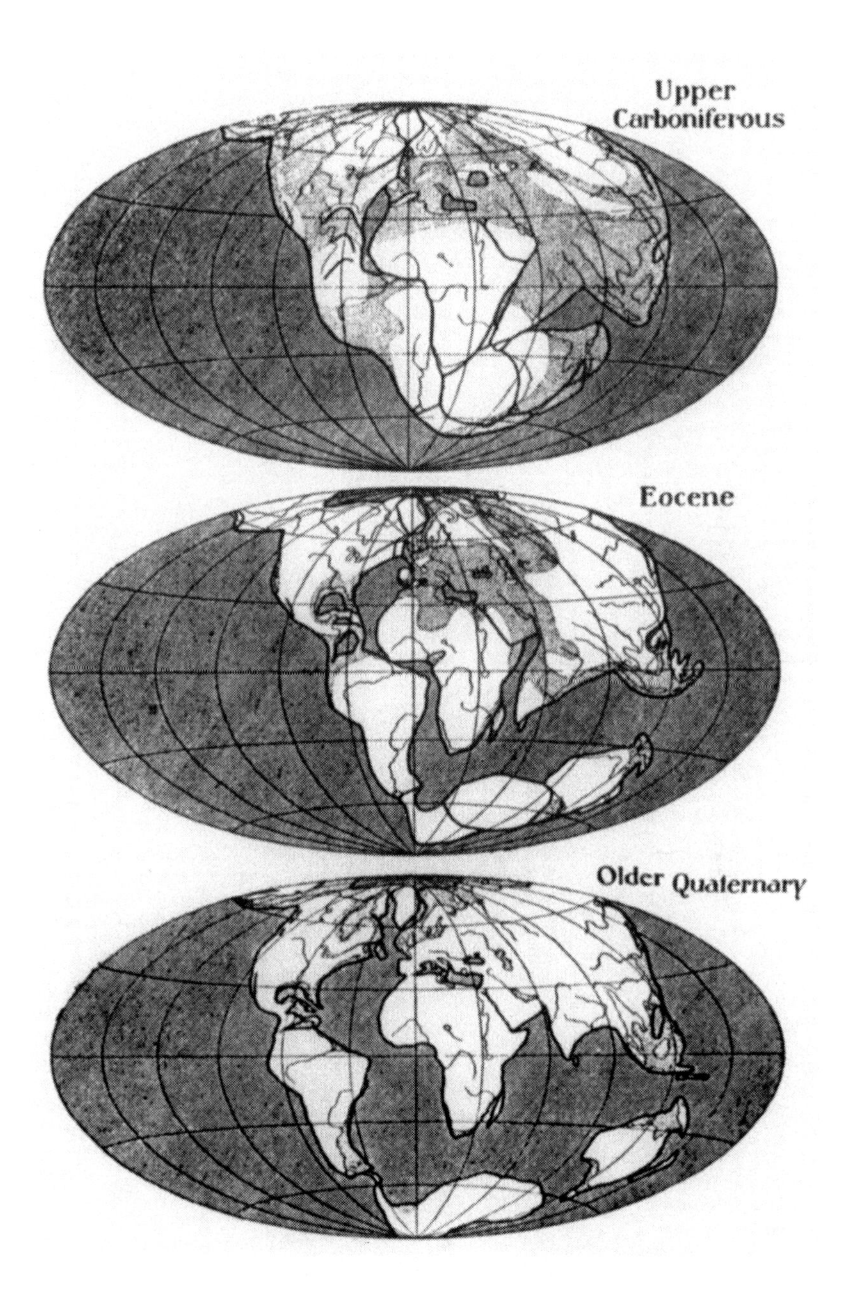

Alfred Wegener's reconstruction of the breakup of Pangaea.

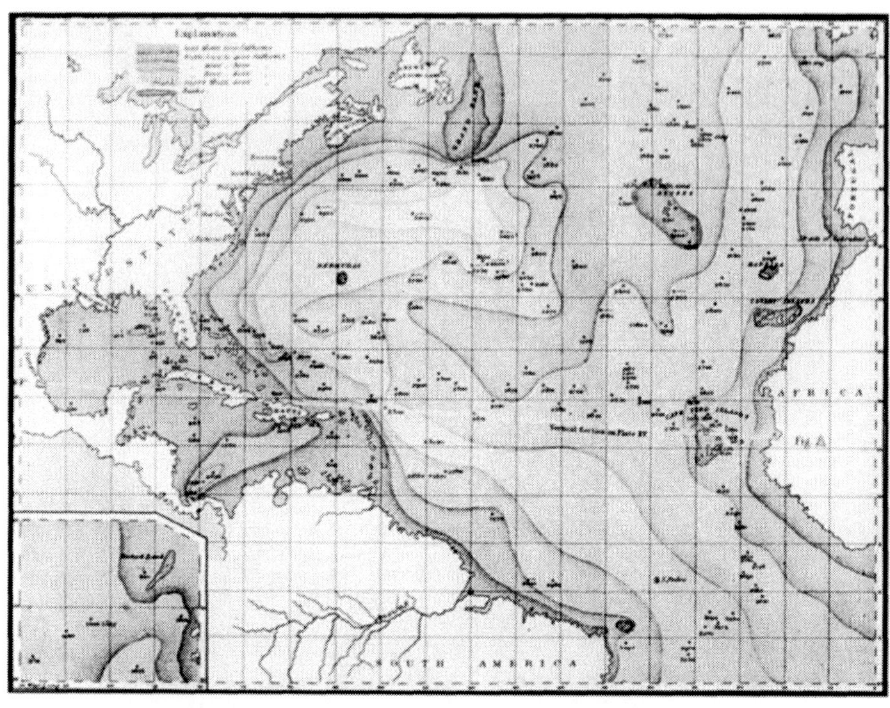

▲ Matthew Fontaine Maury's map of the North Atlantic, ca. 1859.
National Oceanic and Atmospheric Administration Central Library.

▼ Sir John Murray's map of the North Atlantic, 1911.
National Oceanic and Atmospheric Administration Central Library.

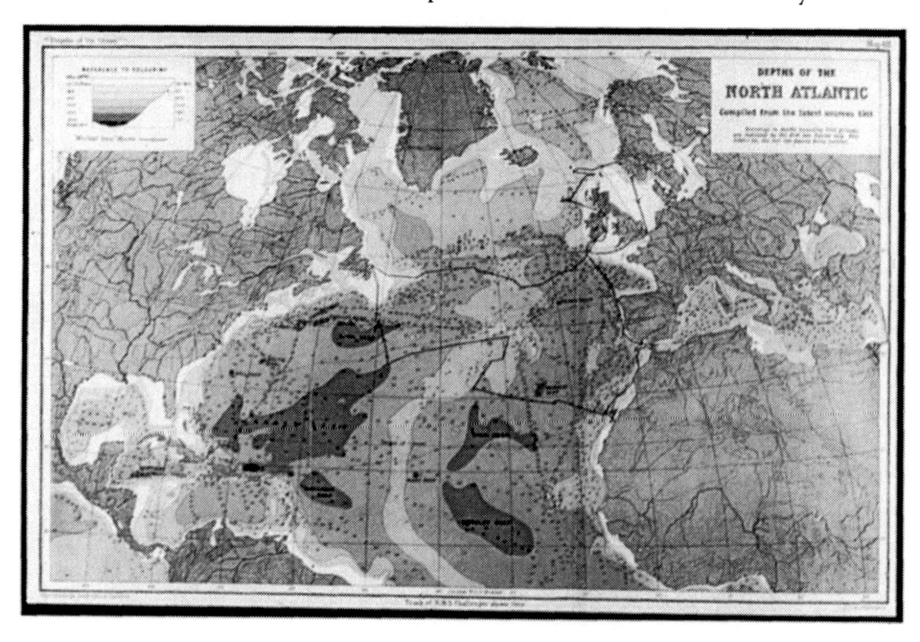

▲ The Fessenden oscillator suspended from boom above the deck of the *Susie D.* National Oceanic and Atmospheric Administration Central Library.

◀ R/V *Atlantis.* ca. 1931. Copyright Woods Hole Oceanographic Institution.

▲ Maurice Ewing in the chief scientist's cabin, R/V *Atlantis,* 1948. Copyright Woods Hole Oceanographic Institution.

▼ Maurice Ewing asleep on deck of R/V *Atlantis.* Copyright Woods Hole Oceanographic Institution.

Seismic shot off stern of R/V *Atlantis.*

▲ USS *Barracuda,* one of the submarines used in conducting gravity surveys in the 1930s. Reproduced by permission of George B. Hess.

▼ Lt. Cmdr. Harry Hess on the deck of the USS *Cape Johnson.* Reproduced by permission of George B. Hess.

▲ Busy waters off Iwo Jima, with Mount Suribachi in background. Reproduced by permission of George B. Hess.

▼ Echo sounder trace of guyot discovered by Harry Hess in the Pacific. National Oceanic and Atmospheric Administration Central Library.

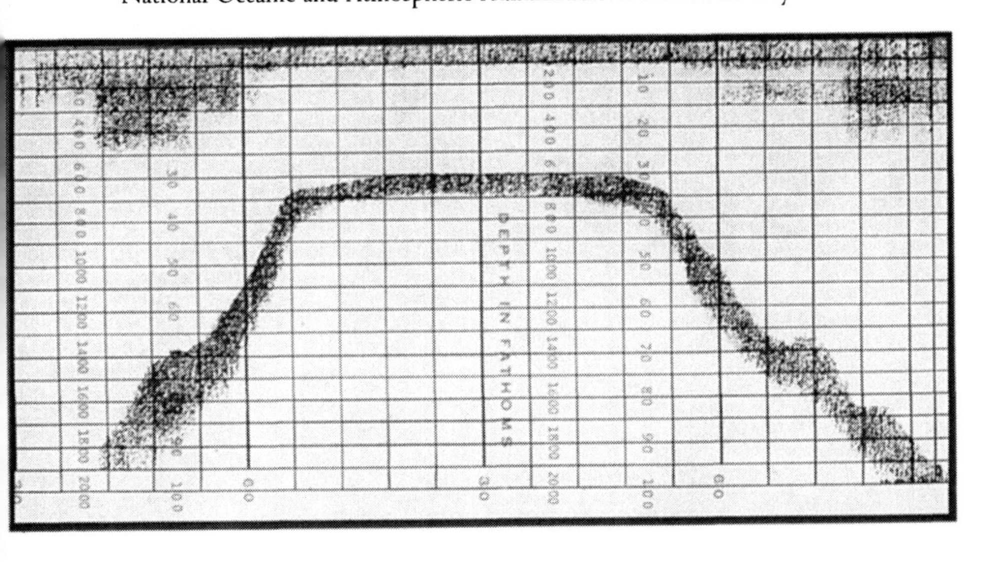

Echo sounder profile of the Mid-Atlantic Ridge.
Copyright Woods Hole Oceanographic Institution.

The young *Vema.* Lamont-Doherty Earth Observatory.

▲ *Vema* on the Hudson River. Lamont-Doherty Earth Observatory.

▼ Marie Tharp in her drafting room. Lamont-Doherty Earth Observatory.

▲ Bruce Heezen, *left,* with Mike Brown. Lamont-Doherty Earth Observatory.

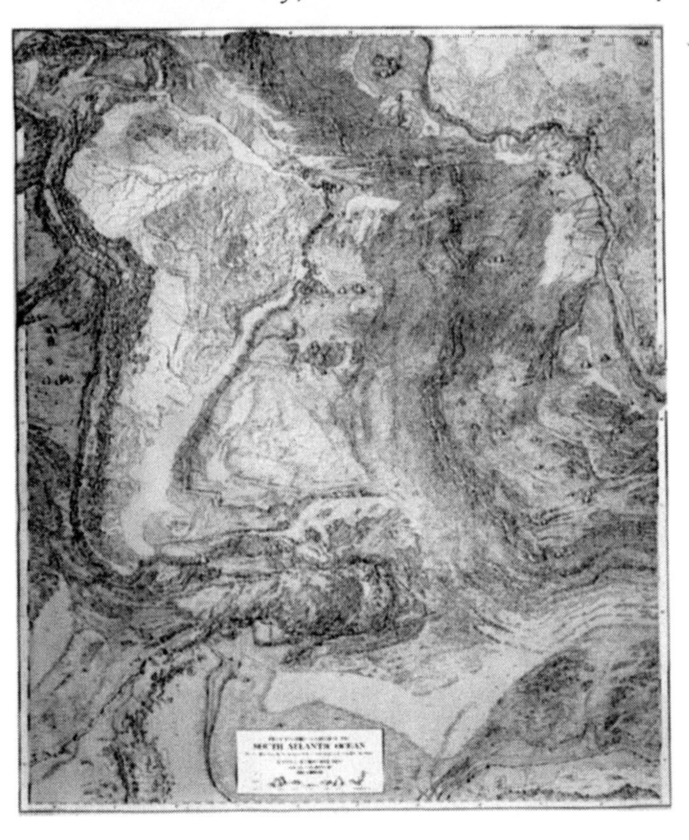

◀ Bruce Heezen and Marie Tharp's map of the South Atlantic. Copyright 1977 by Marie Tharp. Reproduced by permission of Marie Tharp.

▲ Harry Hess in his office at Princeton. Reproduced by permission of George B. Hess.

Harry N. Stewart ▶ checking a ship-towed magnetometer. U.S. Coast and Geodetic Survey.

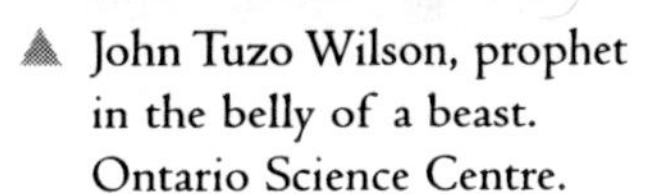

▲ John Tuzo Wilson, prophet in the belly of a beast. Ontario Science Centre.

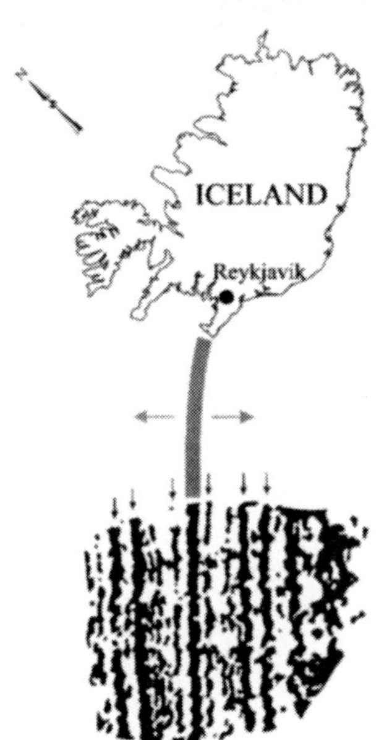

◀ Magnetic profiles along the Reykjanes Ridge. Gray arrows indicate direction of spreading from rift; black arrows indicate matching magnetic patterns (symmetrical on either side of rift).

Some principles of plate tectonics.

(A) Transform fault between two midocean ridges: gray arrows indicate direction of spreading from rift; black arrows indicate movement in opposite directions along either side of transform fault. (B) Seafloor spreading: convection currents in earth's mantle rise under midocean ridge, descend along deep-sea trench; note volcano on opposite side of trench.

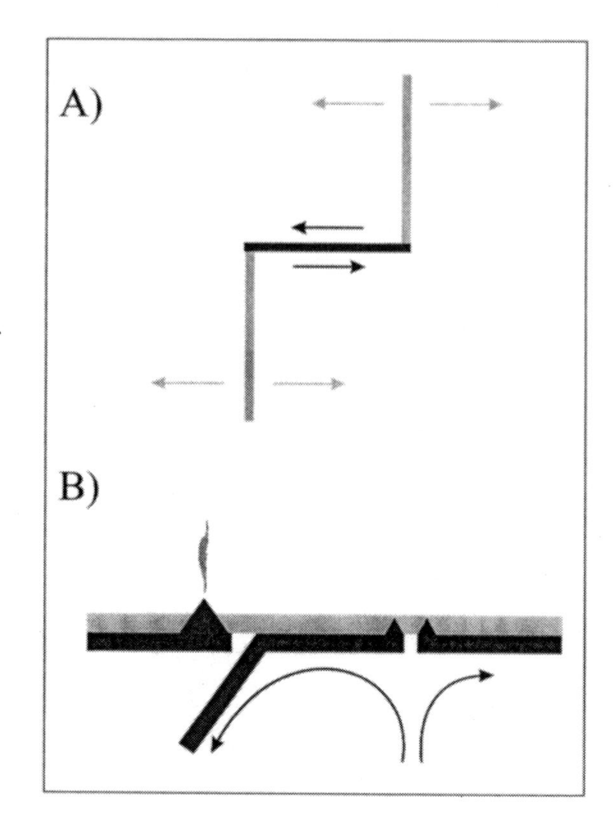

Vema, now the cruise ship *Mandalay*, 1999. Photo by David M. Lawrence.

use on board a submarine. The Ewing and Vine submarine BT had another benefit besides helping the boat crew avoid detection. Buoyancy is affected by temperature and pressure. By knowing the temperature profile in the water column, the submariner would be better able to predict the amount of ballast required to maintain a given depth. Not only were many lives saved by the Ewing and Vine BT, but the tens of thousands of glass slides that were sent back—either to Scripps Institution of Oceanography in La Jolla, California, by vessels operating in the Pacific and Indian Ocean or to WHOI by vessels operating in the Atlantic—enabled oceanographers to learn a great deal about the complex temperature regimes of the world's oceans.

The first ten BTs were built in the machine shop set up by Vine and Worzel. In time Ewing organized the transfer of BT manufacturing to a commercial company, oversaw the design of a circular slide rule to make it easier to calculate sound-ray paths in water, and devised a BT training program for navy personnel that was eventually taken over by the navy's Bureau of Ships.

Between September 1940, when Vine and Worzel returned to Woods Hole, and the following January, the men in Ewing's group had been working long hours—from eight in the morning to ten or twelve at night, with only short breaks for lunch and dinner. They had completed a major report on sound behavior in ocean water and built ten working bathythermographs along with a prototype winch for the BT. Unfortunately, funds for the NDRC had not yet been appropriated. Ewing and company had an impressive list of accomplishments to show for their first four months of wartime research—but they had not been paid. They were finally rewarded with a nice paycheck for their efforts near the end of January 1941.

The Ewing group's expertise with underwater photography also proved useful to the war effort. While experiments to determine whether underwater cameras would be useful in detecting enemy mines proved unsuccessful, the cameras had other uses. A rapidly accumulating number of wrecks littered the sea off the

East Coast, but their positions were not well known. The confusion made navigation in waters less than thirty fathoms deep rather hazardous. Thus the Ewing group was called upon to identify a number of wrecks between Montauk Point on Long Island, New York, and Cape Hatteras. Late in 1942 Ewing and Worzel were dispatched to an unidentified wreck site off the coast of North Carolina. After a week of attempts, a small fleet of ships could not identify the hulk on the bottom. Ewing and Worzel took twenty-two photos in the span of an hour and identified the wreck as *U-352*, which had been scuttled by its commander after it ran aground during a depth-charge attack by the Coast Guard cutter *Icarus* on May 9 of that year. One of the pictures showed the lower portion of a hip boot—apparently left by a crewman abandoning the stricken boat—sticking through the conning tower hatch.

On another occasion a Coast Guard cutter, the *General Green*, and a converted yacht were engaged in an intense battle with a submerged target near Nantucket Shoals, which, "although heavily damaged and making an oil slick, is still moving." Ewing was dispatched with a hastily assembled crew on board the WHOI ship *Anton Dohrn*. He found the cutter and yacht aggressively circling their prey, waiting for it to surface. A marker buoy was dropped from the *Anton Dohrn*, and several photographs were taken. The prey was in fact an old wreck of the Nantucket Lightship! "The attack ended there," Ewing said. "In breaking off, the cutter got up to full speed, spent all her remaining depth charges in a salvo at the spot marked by our buoy, then asked us to lead her to Gay Head Light, reporting that the last salvo had knocked out all her compasses."

In the fall of 1943 Ewing's group was called upon to determine the behavior of sound between ten and twenty fathoms. The request covered all sound frequencies in a wide variety of bottom and geological characteristics. When Ewing sought more information about what the navy wanted to learn, he found out that his group did not have any "need to know." He guessed that the

Germans were destroying entire fields of acoustic mines by detonating a single charge at a safe distance from where the mines were laid.

Doc and company selected five areas in which to conduct acoustic tests for the project: Solomons Island in the Chesapeake Bay for thick mud; the Atlantic off Jacksonville, Florida, for thick sand; the Virgin Islands for volcanic rock; Barbados for coral over sand; and near Trinidad for a thick mix of sand and mud. The experiment basically consisted of seismic refraction work. A Coast Guard cutter would deploy the charge—and provide protection against German submarines—while the *Saluda,* a converted yacht, would deploy geophones and hydrophones to record the sound waves. Worzel was on the cutter dropping the bombs while Ewing was on the *Saluda* recording the sound arrivals at various distances.

While off Jacksonville, sonar on the cutter detected a submerged target that was moving. The captain radioed a nearby Coast Guard base that a submarine was in the area and began attacking with depth charges. Despite repeated pleas for assistance no help came, and the cutter broke off the attack near sundown to return to the defenseless *Saluda.* The next day a freighter was torpedoed about thirty miles north of where the cutter and the *Saluda* were working.

Near Trinidad the Coast Guard cutter was replaced by a subchaser. At one point in the experiment Worzel had to use depth charges, but the first three dropped did not go off. When trying to determine what was wrong with them, he learned that they had been reworked by a navy Bureau of Ordnance installation in Trinidad.

(The Bureau of Ordnance had a dismal record before and during World War II—sending ships out with many defective weapons and being reluctant to correct the problems. The submarine fleet was the most adversely affected by the department's incompetence. Their weapons of choice, torpedoes, tended to run too deep. Torpedoes equipped with magnetic exploders—which were supposed to detonate when reaching the magnetic field of a target's hull—were sensitive to magnetic fluctuations in the environment, with premature explosions being the unfortunate result. The firing mechanism on torpedoes equipped with contact exploders were also defective. The contact exploder functioned well when striking a target at an angle, but the firing mechanism was often crushed by the impact of a straight-on shot, and thus the warhead would not detonate. The bureau generally ignored the submariner's complaints until controlled tests—by the submarine commands themselves rather than the bureau—proved the gripes were well founded.)

When told that the depth charges had been reworked, Worzel deduced that they must have been assembled improperly and asked to see if any of the engineering drawings were on board. The subchaser's ammunition officer balked—after all, how could a civilian question the work of the Bureau of Ordnance—but the captain overruled him. Worzel studied the drawings for about an hour; then he and one of the sailors examined the detonator mechanism and identified the problem. The repaired depth charge was reassembled and tested. It exploded as it should. The captain ordered the remainder of his supply of depth charges repaired. The captain may have been grateful—since he did not have to return to base, reload with functional depth charges, and return to complete the mission—but the ammunition officer refused to speak to Worzel for the rest of the cruise.

While working the waters near Trinidad, the subchaser and *Saluda* were hit by an intense gale. As the storm worsened, Worzel lashed himself to deck within reach of the explosives and kept working. Finally, the captain asked if he should quit until the storm abated. Worzel answered that as long as Ewing could keep manning the listening equipment on the *Saluda,* he could keep firing the shots. Doc was asked a similar question by the captain of the *Saluda.* His answer was that as long as Worzel could continue dropping bombs, he could continue recording. After three days in the storm, they completed the work.

What Ewing and Worzel learned from the series of tests is that sound waves disperse in shallow water. They also demonstrated that low-frequency waves traveling through the seafloor, even though they must travel a greater distance, reach a listening point first, while higher-frequency waves traveling through the water itself then follow. Acoustic mines would explode from the low-frequency pulses before the high-frequency waves traveling through the water could make them inoperative, as they were supposed to. The navy adjusted the firing mechanisms of the mines to make them less vulnerable to sabotage.

The results confirmed an untested theory on sound transmission in shallow water by Chaim Leib Pekeris, a professor at Columbia University, and eventually led to an explanation of the dispersion of earthquake surface waves by Ewing and Frank Press after the war.

Gunfire and Geology

Harry Hammond Hess experienced a different kind of war. His job was not research but fighting.

Hess—the son of Julian S. Hess and Elizabeth Engel Hess of New York City—was born twelve days after Ewing. Unlike Ewing, however, he came from a fairly well-to-do background. His father was a member of the New York Stock Exchange. His paternal grandfather, Simon Hess, operated a construction firm based in New York, and his maternal grandfather, Julius Engel, ran a distillery in Hammond, Indiana.

After graduating from Asbury Park (New Jersey) High School, Hess attended Yale University in New Haven, Connecticut. He initially majored in electrical engineering but wearied of having to draw endless schematics of gadgets like spark plugs. He transferred into geology—where Hess said he flunked his first course in mineralogy and was warned by his professor, Adolph Knopf, that he had no future in the field. Hess ignored the warning and graduated from Yale in 1927.

He accepted a job with Anglo American mining in Northern Rhodesia (now Zambia). He spent two years in Africa, doing geological surveys on foot in the southern African savannas, wearing a pith helmet and shorts, sleeping in tents, and using African porters to haul the gear. He found time to take plenty of photographs as well as produce finely detailed sketches of his experiences. The time in Africa proved to be an education of a different sort than what he had received at Yale. "At 17 miles a day, I developed leg muscles, a philosophical attitude toward life, and a profound respect for field work," is how he later summed up the experience. Hess had indeed learned how to adapt to exotic conditions, to work with people of completely different backgrounds, and to take charge in a quiet, unpretentious way.

Hess returned to the United States in 1929 with a desire to go to graduate school. He first considered Harvard, but the chain-smoker was put off by No Smoking signs around the buildings on the campus. He then headed south to his alma mater, Yale, but was told that the university had seen more than enough of him as it was. Princeton was suggested as an alternative.

Upon his arrival in Princeton, Hess tracked down Arthur Buddington, who confused him with another Harry—Harry Cannon—to whom he was going to offer a part-time position as an instructor. As they were filling out the paperwork, Buddington was horrified to learn that he had the wrong man. At that moment Richard Field, the man who had introduced Maurice Ewing to ocean-floor research, walked in and saved the day for Hess.

"I had been on [Field's] summer school on a railroad car in Canada two years before," Hess later recalled. "On the last night before coming back into the United States he told us all liquor on the car had to be disposed of before the border was reached. It was during prohibition in the United States. I consumed far more than my share. Egged on by my fellow students, I was put on the platform and gave a lecture on the Precambrian stratigraphy of Canada. I have no recollection of this lecture but am told it was good."

Field remembered the lecture and insisted that Hess could do the job. Hess was allowed to stay. He graduated with a doctoral degree in 1932 and joined the geology faculty at Princeton two years later. He would be affiliated with the school for the rest of his life.

FIELD MANAGED to steer Harry Hess toward ocean-floor research—as he had done with Ewing and Sir Edward Bullard. Field was an early American supporter of the work of Felix Vening Meinesz and, while arranging passage for the Dutchman aboard a navy sub for a research cruise to the West Indies and Caribbean, talked Hess into going as well. Hess and Vening Meinesz worked together on the *S-48* cruise in 1932. Hess later accompanied Ewing on the *Barracuda* cruise from Panama through the West Indies to Philadelphia in 1937. The experience of the *S-48* cruise may have played an influential role in Hess's decision to apply for a commission in the Naval Reserve in 1934. He received his commission—as a lieutenant, junior grade—early in 1935.

As a result of his commission, Lieutenant (jg) Hess had a job waiting for him after the Japanese attacked Pearl Harbor. The next day he boarded the 7:42 A.M. train in Princeton, bound for New York City to report for active duty. The navy did not take any time to evaluate his credentials and see how he could best fit in with the war effort. When he arrived at the Eastern Sea Frontier headquarters at 90 Church Street in New York, he was asked if he knew about longitude and latitude. He did, and was promptly given a piece of chalk, pointed toward a blackboard, and told to plot reported U-boat locations. Hess did that for a time, but soon he spotted a chief petty officer walking around and asked, "Do you know about longitude and latitude?"

Before long Hess was given the responsibility of predicting where U-boats would be. German Grossadmiral Karl Dönitz made Hess's job easier. Dönitz insisted that his boats radio reports to headquarters on a daily basis. As they did, Allied listening posts could pinpoint their locations—and Hess had a knack for guessing where the boats would be the next day. His superiors were impressed with the "kill" rate achieved on the basis of Hess's recommendations. He did not gloat, however, but modestly claimed that his geological training prepared him well for determining patterns from sparse data.

Hess wanted to go to sea, but Vice Admiral Adolphus Andrews, commander of the Eastern Sea Frontier, was reluctant to let such a good sub hunter go. Eventually Hess did get to go to sea, first on the mystery ship *Big Horn*, a converted tanker designed to lure, and hopefully kill, U-boats careless enough to attack it. However, between September 1942, when the *Big Horn* went to sea, and early 1944, when the navy gave up on the experiment, no German submarine took the bait.

Commander Louis C. Farley, who had been the skipper of the *Big Horn*, and much of his crew, including Hess, were next transferred to the *Cape Johnson*, a freighter built for the Maritime Commission that was being converted into a troop transport at Los Angeles Shipbuilding and Dry Dock Company in California. While at the shipyard, Farley ensured that the *Cape Johnson* was supplied with twice the planned refrigerator capacity and that the freezer was well stocked with ice cream to keep the sailors and marines or soldiers in transit happy. Unfortunately, the sick bay door wasn't large enough to allow a stretcher to pass; a hole had to be cut in the deck when the first casualties came in so that their stretchers could be lowered into the bay.

Hess began as navigation officer—third in command—on the *Cape Johnson* when it departed from Los Angeles bound for the Marianas in the summer of 1944. He had a piece of equipment on board that he grew to value highly: a fathometer. He kept the fathometer running—and collecting sounding data—almost constantly. Soon Hess was promoted from navigation officer to executive officer.

Cape Johnson spent its first four months of duty participating in the assault on Tinian in the Mariana Islands and transporting marines and soldiers from islands where they were no longer needed to bases where preparations were being made for further amphibious operations. Early in November 1944 the ship departed Hollandia, New Guinea, on a mission to resupply forces participating in the invasion of Leyte Gulf in the Philippines. Along the way, on the thirteenth, the transport sailed unscathed through an intense Japanese air attack on the convoy and safely put its passengers ashore at Samar two days later.

The transport then sailed to Manus, in the Admiralty Islands, to prepare for the invasion of Lingayen Gulf on the Philippine main island of Luzon. *Cape Johnson* weathered two days of intense air attacks before landing its men on White Beach, near the town of San Fabian, on January 9, 1945.

Cape Johnson spent the next few weeks at Guam, in the Mariana Islands, and at Ulithi, in the Caroline Islands, training for the next big operation—the assault on Iwo Jima. Hess took advantage of the relative break from action to write a student of his, who was now Lieutenant (jg) John C. Maxwell, also stationed in the Pacific. The first two paragraphs of the letter, dated February 2, are a snapshot of navy life in the remote reaches of the Pacific.

> Dear Mac
>
> We are having a short breather between trips, anchored in one of those atolls where going ashore in our spare time offers little or no attraction and a long hot boat ride. . . .
>
> We've travelled a lot almost entirely from out of the way advanced bases to new beachheads. I think I wrote you when we were in Espiritu last fall. That was my first trip as Exec and that is what I am now, still harassed by messengers and paper work but getting more or less used to it. I just discovered that I was promoted a couple of months ago. . . .

Two weeks later the *Cape Johnson* left Guam for Iwo Jima. Hess recorded the events of the assault in his navigator's diary:

> Departed Guam with Marines of 3d div. embarked late afternoon of Feb 16. We are now temporarily in Transport Division 32. Our old 6th division having been broken up after the Lingayen Gulf operations last month.
>
> We are carrying reserves for this operation. Fourth and fifth M.C. divisions make the assault landing and we wait just off shore, standing by for call if needed.
>
> Feb 17 and 18 uneventful sailing. One false alarm air alert noon the 18th. Many of our own planes in air.

The pre-invasion bombardment of Iwo Jima began on the seventeenth. Hess continued:

> Evening of 18th wind rising and heavy swell—rather unfavorable for landing tomorrow unless weather moderates. Reports from bombardment task group not too favorable either. Many known emplacements not yet knocked out. Japs holding their fire so that gun emplacements are not given away before assault phase.
>
> Feb 19, D Day: It has been a quiet night. The sea and wind have died down, conditions are excellent for landing. Iwo Jima is in sight. Radio just reported that the first assault was now fifty yards from the line of departure.
>
> 0900: first wave hit beach. Little if any enemy fire observed by Recon plane up to this point.
>
> One of our two Recon planes shot down, crashed in water just off beach.
>
> 0930: Troops about 200 yds inland over first terrace. Some scattered mortar fire on beachhead.
>
> 0940: We have gone to condition 1A and seem to be maneuvering to take station at rear of transport area.
>
> The temperature is near 70° F. I was shivering and had to [put on] my winter coat. . . . We are not used to anything below 80°.
>
> 1200: We seem to have a beachhead about 500 yds deep along whole right flank. Data on left flank not known but some tanks are on SW end of airfield so penetration seems to be deeper here. Enemy concentration right of our right flank. Counterattack may be expected from this area tonight. Our troops will be used to reinforce right flank if needed.
>
> Naval gunfire being called for on various scattered targets as well as plane strikes. Everything quite peaceful out here in the transport area. Can see our tanks moving about across beachhead and returning to be rearmed. It looks as though we would not land our troops today.
>
> Recon plane reporting progress of troops was talking rapidly and somewhat excitedly. HQ ship told him to be calm. He said "Junior" was shooting at him. Next thing we knew he was shot down. Don't know if he was recovered—hit water right off beach.
>
> About 1600 small shells began popping around us as they hit the water. Sent troops below.

At dark started to retire with Divs 43, 44 and 45. Much confusion in forming up in smoke and dark. Progress of attack can only be guessed at from out where we were watching. Opposition appeared to be fairly heavy and well organized. Much of our artillery had not been landed. Amount of ammunition ashore for what was landed not large. Mortar fire in central part of our beach area destroyed some dumps and interfered with landings. Our rocket planes seemed to be very effective. Doubt if troops reached objectives for D day. May be in some trouble if there is a heavy counterattack tonight.

No big counterattack was launched that night. Lieutenant General Tadamichi Kuribayashi did not plan to waste his men in one big banzai charge—as other Japanese commanders had repeatedly done throughout the war. Kuribayashi dug in and intended to make the marines pay dearly for his particular piece of real estate.

Feb 20: Returned to beachhead. Before dawn very heavy fire.

0800: Put our boats in water. Sent platoon to the *Jupiter* and seven boats to the *Pres. Adams.* 21 RCT went into boats (our troops did not disembark today) approached the blue beaches but held off shore due to mortar fire. 1500: these troops reembarked on transports. We should retire again tonight. It is raining and visibility is poor.

Things seem to be going all right on shore. This morning the lower third of the island was secured except for the volcano at the southern tip. Fourth and Fifth divs. slowly moving northward. There can be no doubt of the outcome now. Though there will still be some hard fighting over the rough terrane [*sic*] to the north of our present lines. *APA 196* collided with *APA 157* during night retirement.

21st—Weather bad, heavy swell, strong wind. Very difficult to put boats in water. Got all our boats over in about one hour with no casualties. They went to *Pres. Adams* and assisted in putting 21st RCT ashore. Lost one boat on the beach VP#3. There was sporadic mortar fire along beach. One boat broached. We were not permitted to attempt salvage. Lost another VP trying to hoist boats aboard later in day.

Resistance on island heavy. The commanding general [Lieutenant General Holland M.—"Howlin' mad"—] (Smith) says it is heaviest ever encountered by marines. Five thousand additional [Japanese] troops on island which were not previously known of.

23d: Standing by again on 22d. Rain and heavy swell. Towards evening troops on shore were running short of 81 mm mortar ammunition. Stayed behind to unload 81 mm ammo, and send it in. We got ours off at 2100 in one of our own LCM—Lt. jg Kemp in charge—and one of the *Pres. Adams* LCMs. Had a much disturbed night maneuvering off island in strong wind and heavy swell. Many landing craft lost in area. At 0300 heard our LCM was in distress 6 miles off south end of island bearing 125 approx. Searched but could not locate it. Concerned over possibility it will drift rapidly to sea.

At 0500 sighted three men in water. Turned search light on and got whale boat away. Picked up the men from *LST 764*. Their LCVP had been swamped and sank. All of crew picked up. Put them in sick bay.

At sunrise hoisted two more LCMs into water with great difficulty. These were to carry ammunition from AEs to beach—Ens. Peddicard in charge. Div. Commander gave us a "Well done" for getting them away.

About 1030 *PC 800* towed our missing LCM alongside. They had delivered ammo. O.K. but hit submerged LCVP wreck off beach. Fouled both screws, only one engine would turn over. Managed to get off beach however by skillful handling. Lost one man overboard but he was recovered by a duck and returned to boat. Boat had too little power to make headway against seas. Drifted 14 miles to SE but fortunately was picked up by inbound convoy escort.

Jap planes coming down more regularly now. Saw one bomb splash in water during squall afternoon of 22d. Tonight I was in CIC after sunset. Many bogies on screen. Considerable AA fire from convoy to north of us. One explosion near ships to south. Don't know whether plane exploded or ship was hit. We had no fighter cover after dark. I think we are in for trouble in the next few days, particularly at dusk or immediately after.

Our lines on island stationary for last 48 hours or practically so. Marines reforming. 21st RCT or 3rd Marine Div. replaced a 4th Marine RCT in center of line. Only half of our troops (ship's) are shore now and none of cargo except 81 mm ammo. I guess we'll be around here for another ten days if we are to wait to take the marines

> back. Over 4000 casualties so far. Percentage killed very low. Mt. "Hotrocks" volcano has our flag flying from top of it since 1030 today. Some fighting, however, still going on in and around it.

Mount Hotrocks was the code name for Mount Suribachi, the highest point on the island. Marine First Lieutenant Harold G. Schrier led a small detachment of the Twenty-eighth Marine RCT to the top of the volcano. Schrier, Sergeants Henry O. Hansen and Ernest I. Thomas, Corporal Charles W. Lindberg, and Privates James R. Michaels and Louis Charlo raised the first American flag on the summit at about 10:20 in the morning. The better-known raising of the second flag—captured in a Pulitzer Prize–winning photograph by Associated Press photographer Joe Rosenthal—occurred about fifteen minutes later.

> 24th: Came in and anchored. Got some more troops ashore. About 1000 a TBF was hit and caught fire, she flew out over the water at about 1000 feet. Pilot (and perhaps one other) bailed out. Parachutes did not open. Plane crashed almost into *Pres. Jackson* on our starboard bow.
>
> Damaged another LCVP trying to hoist it aboard. Four men had to jump into water. One slightly injured where hit by light attached to life ring which was thrown to him. All were recovered.
>
> 25th: Came in very close to the beach this morning opposite Jap held east or north east side. Shots from our cruisers were falling all around us so we had to move out even though they were "friendly" shells.
>
> Had a very clear view of the marines advancing on airfield number #2 [*sic*] yesterday afternoon and again this afternoon. Most of troops now ashore and ammunition but 90% of cargo left.

The *Cape Johnson* spent five weeks and a day off Iwo Jima, landing men and equipment when needed and helping to fend off Japanese air attacks. When the ship finally departed on March 27 carrying men of the Fifth Marines, it was the only ship left that had been present on D day.

EVEN WHILE under fire, Harry Hess thought about geology. He wrote Maxwell again on March 22, five days before leaving Iwo

Jima. In a series of letters, Hess and Maxwell had been discussing the tectonics of New Caledonia. Now Hess tried to make some sense of some of Maxwell's ideas:

> Dear Mac:
>
> We went up with the assault on Iwo Jima so this is about the first chance I've had since my short note to finish answering your letter of Nov. 15.
>
> Since you have proven the intrusive nature of the serpentines in those limestones and sandstones all that remains is to check the forams to determine the age of the [rocks]. It seems hardly likely that they could be very far in error and yet an Eocene age for the serpentines raises many perplexing questions and also seems unlikely. So I'm stumped. . . .

In the next few paragraphs of the letter, Hess discussed the pros and cons of several possible ages of the mysterious serpentine rocks. He concluded the passage with a paragraph that reveals much about his thought processes.

> So you see I'm trying to make the geology of a large area fit my preconceived opinion as to what it ought to be—and this even though I've never seen 90% of it. I'm aware that I am doing this, but would persist in it because I believe that regional geology always makes sense and fits together when you've got the story right. It doesn't fit at all now so in spite of the apparent facts I'm going to hold on for a while to my unwarranted hypotheses, and look upon the "facts" with suspicion.

Hess went on to discuss the research he had managed to accomplish so far as well as life as an executive officer.

> My own investigations are rather difficult to write about. Just a large collection of soundings and rocks when I can get ashore for a few minutes. Have been able to get about a dozen or more traverses across deeps and can outline their course pretty well from Iwo to Palau. Have four across the Mindanao deep, too. We filled in a lot of blank spots on the charts.
>
> I'm getting used to being Executive Officer. Having delegated most of the nasty jobs to subordinates, I keep almost entirely out of

> the Ship's Office now. Everything seems to be running very smoothly right now.

Hess, however, did not get to enjoy being executive officer for long. In Eniwetok on April 3, he was informed he was now the *Cape Johnson*'s commander.

BY WAR'S END, Hess had seen a lot of the ocean. His experience wasn't limited to the waters and lands at the surface but also—thanks to the fathometer—included the features of the depths. In addition to charting numerous deeps, Hess had discovered a unique structure never before imagined: flat-topped, submerged mountains that he called guyots in honor of nineteenth-century Princeton geologist Arnold Henri Guyot. Fifteen years after the Japanese surrender, Hess would use his knowledge of the ocean floor to launch his own assault on the earth science establishment.

The Upstarts

Maurice Ewing did not return to Lehigh after the war. Instead, he accepted in 1944 a position as an associate professor of geology at Columbia University. In February 1946 he set up an office in Schermerhorn Hall Extension—on Columbia's main campus in the Morningside Heights section of New York City—while waiting for basement rooms to be refurbished in adjoining Schermerhorn Hall. The basement offices had recently been vacated by the Manhattan Project—the United States's effort to build an atomic bomb. (During the Manhattan Project, Enrico Fermi assembled, in Schermerhorn Hall, the first experimental nuclear pile. Members of the Columbia football team provided much of the heavy lifting for the work.)

In the summer of 1946 Ewing returned to Woods Hole with two of his graduate students, Nelson Steenland and Frank Press. Joe Worzel joined the group, and they shot a number of seismic refraction profiles across the continental shelf using *P-446*, an air-sea rescue vessel on loan from the United States Army Air Force, and the *Balanus*, a member of WHOI's fleet. Gordon Hamilton, another graduate student of Ewing's, joined the group later that summer. Ewing trained Hamilton, Steenland, and Worzel to make pendulum-based gravity measurements while on board the navy submarine *Tusk*. The *Tusk* conducted a gravity survey along the same transects that the *P-446* and *Balanus* had traveled earlier.

In the fall Ewing's entourage—except for Allyn Vine, who remained at Woods Hole—took up residence in Schermerhorn Hall.

Ewing resumed his seismic work in the summer of 1947. Presented with a wonderful gift from Woods Hole—sole use of the *Atlantis* for two months—Doc intended to use new devices and techniques developed during the war to intensively study an area long ignored. The area was the Mid-Atlantic Ridge.

Scientists had known for almost a century that there was some kind of rise in the North Atlantic. Matthew Fontaine Maury, working with only two hundred widely scattered soundings, marked it as a plateau on his 1854 map of the North Atlantic floor. Charles Wyville Thomson, on the *Challenger* expedition, used a more dense network of soundings and temperature measurements to confirm that a broad rise lay in the middle of the North Atlantic. Sounding and temperature data from the *Meteor* expedition showed, however, that the rise was not broad and gentle but narrow and extremely rugged and that it extended into the South Atlantic. Close examination of the *Meteor*'s sounding data would also have revealed something else about the ridge—something that could have blown the lid off the debate over Alfred Wegener's theory of continental drift—but the rise of the Nazis and outbreak of World War II interrupted analysis of those data.

Ewing and crew were frantically getting *Atlantis* ready to sail on July 16, 1947. Hydrophones and a ton of TNT were being loaded for the seismic work. Ewing had learned during the war to treat water as just another layer on the surface of the earth rather than a substance that obscures the rocks beneath. He realized it was unnecessary to lower geophones and bombs to the solid bottom: The insight made seismic research at sea much easier to carry out. Deep-sea cameras and coring and dredging equipment were also taken aboard to enable Ewing to better characterize the ocean floor.

Columbus Iselin, in an unpublished memoir, commented on Ewing's influence on WHOI. "He had a profound effect on the success of this laboratory. He arrived here as a very young professor," Iselin wrote. "He brought with him several Lehigh students, and the place has never been the same since. They literally worked night and day and seven days a week."

Ewing was certainly reinforcing that impression on the Woods Hole dock that day. "We're paying for this ship twenty-four hours a day," he said loud enough so many could hear, "so damn it, we're going to *work* twenty-four hours a day."

Iselin may have felt inspired by Ewing's influence, but his opinion wasn't shared by all. One who did not, unfortunately, happened to be the ship's cook—who decided to quit less than an hour before sailing time. *Atlantis*'s captain, Adrian Lane, was trapped in the maw of a major crisis—he could not do without a cook for a two-month cruise. Lane's attempt to negotiate with the obstinate man in the crew's quarters was unsuccessful. In desperation, the captain suggested that the cook leave his gear on his bunk while they went across the street to Cap'n Kidd bar. The cook yielded to rum and reason—or a "reasonable" facsimile thereof—and eventually accompanied Lane back on board.

The *Atlantis* initially sailed to Bermuda to take on additional supplies. Little of importance happened on that leg of the voyage. The real work began after the ship left Bermuda on July 24. The echo sounder picked up a small seamount east of the island, and Ewing altered the ship's course to survey it.

EWING KEPT the echo sounder on the *Atlantis* running constantly while the ketch was at sea. The *Atlantis*'s sounder was a greatly improved instrument over what had been available prior to World War II. It would send out an electronic ping at a regular interval, and a microphone inside the hull of the ship would pick up the echo from the bottom. As the ping was sent out, a stylus would be set in motion downward across a continuously spooled strip of four-inch-wide paper. When the echo returned the stylus would mark the paper by burning it with an electric spark, thus recording the depth much as an electrocardiogram records heartbeats. The resulting trace was a profile of the ocean floor along the ship's course. The sounder recorded depths at a scale of two inches to 1,000 fathoms, or 6,000 feet. Thus, the strips of paper on the echo sounder could originally accommodate depths down to 2,000 fathoms. Characteristically, Ewing asked Worzel to modify the recorder so that it could, by a flick of a switch, also record depths on a scale from 2,000 to 4,000 fathoms.

The sounder on board the *Atlantis* had one major flaw—it was entirely dependent on the ship's overdrawn electrical systems. If someone opened the refrigerator, for example, the electricity often went out and no echo would be recorded. Thus, as far as the sounder was concerned, bottomless pits were found every time the cook opened the refrigerator. Power surges—not uncommon with the ship's balky circuitry—would result in the discovery of ridges. The artifacts were not always immediately noticed. One published "discovery" was later withdrawn—but given the name "Sixty-Cycle Ridge."

If all the crew had to do was to make sure the stylus continued to work and the spooler never ran out of paper, the job at sea would have been easy. But there was much more to it. A scientist held watch in the lab whenever the sounder was operating. With Ewing in charge, that meant someone was on duty twenty-four hours a day, each person typically on eight-hour watches with eight hours off. In the case of the *Atlantis*, the scientist on watch would usually be sitting in a lab below deck, known as the lower lab—space on the deck was called the upper lab—because there was little room on the ship. The cramped, cluttered room often had no portholes or windows, so he (only men were allowed to go to sea on a Ewing ship) could not look out and see the comforting horizon as his stomach turned with the pitching and rolling of a small ship in a big ocean—and *Atlantis* was a prodigious roller.

On most cruises the crew would adjust to the motions of the ship after two or three days at sea. Some scientists, though, repeatedly lost the battle to control their stomach, running to the rail several times on any given watch to throw up. Most of the time, however, the scientist on watch was trapped, swaying with the motion of the ship and watching the recorder, marking down time intervals and course changes on the strip of recording paper, making sure the equipment kept working, and making especially sure that the crew didn't overlook any feature on the ocean floor that deserved detailed study.

At times the lab would be frenetic with crowded, adrenaline-driven activity. Someone might be repairing equipment, analyzing a sediment core, or watching the instruments, keeping awake with coffee from the galley. A group might be arguing over the interpretation of newly acquired data. Some might be relating to others the news from home. But no one would be talking about something silly, such as what kind of chance the Brooklyn Dodgers might have against the New York Yankees in the World Series—for time on ships was at a premium, and everyone had to concentrate on getting as much data as possible. Ewing, or whoever else was serving as chief scientist, might walk through, making the rounds of the research groups to learn how they thought the cruise was progressing. At other times, usually late nights and early mornings, the lab would be quiet, the scientist on watch being the only one awake. Ewing—who rarely slept more than an hour or two at a stretch while on a cruise—frequently stole catnaps on an unoccupied bench, ready to come alive if awakened by someone with a problem or a potentially interesting discovery. If he did sleep in his stateroom, he expected to be awakened if something came up. Anyone who tried to be nice and let him sleep would be rewarded with anger rather than gratitude.

Over the abyssal plains, the ocean floor was as flat and boring as the plains of West Texas are to a long-haul trucker. The depth measurements changed little for interminably long periods of time, leaving the scientist on watch with little to do but mark down times on the recording strips in military style—2300, 2315,

2330, and so on. But eventually the stylus would mark the paper a little higher. The next time the mark would be higher still. The scientist grew more alert, waiting for the next echo to tell him if this was just a slight rise or striking ocean-floor topography worthy of more investigation. A seamount, for example, is defined as a feature that rises at least five hundred fathoms above the surrounding plain. Sometimes it would be obvious that the feature was a seamount and warranted detailed study. Other times the rise would come up somewhat—about three hundred fathoms, for example—before dropping again. Either the feature wasn't a seamount and wasn't worth spending much time on, or it was the flank of a seamount that deserved more attention. The scientist on watch would consult with the officer on duty. They often decided to backtrack across the area on a slightly different course to get a more detailed view of the feature. If nothing more was found, the ship would return to its original course. Otherwise they would call the chief scientist to plan a detailed survey.

While the scientists worried about the depths, the ship's captain and first mate worried about navigation, making sure the course was correctly plotted on their nautical charts. Once a day, one of the scientists would take the depth records from the continuous echo sounder and plot the data on an overlay of the navigational charts, usually working at a scale of 1:1,000,000 (one inch on the map represents one million inches, or fourteen nautical miles, on the ocean floor). Rather than plot every sounding, they plotted the depths only at peaks, troughs, and changes in slope angle along the ocean floor, as well as course changes of the ship. Despite the simplification, there were hundreds of measurements to transcribe each day. They did the work by hand. Despite the motions of the ship, they read the depths off the profiles, with twelve thousand feet compressed to a four-inch-wide strip, and wrote them down on the overlay in neat handwriting so precise it often looked as if it had been typed. They knew an illegible number did no one any good. Hundreds of measurements a day would be transcribed in this fashion.

The scientists watching the echo sounder weren't the only ones keeping busy. Initially Ewing had to stop the *Atlantis* for each seismic shot. While on the cruise, however, a new technique was developed by Doc and Press that eliminated the need to stop the ship. The hydrophones were kept in tow, but just before the shot was to go off one of the scientists would pay out the tow line so that the phones in effect floated still in the water—while the ship kept moving. The new technique was tested on August 14 and proved a success. Now seismic shots could be taken on a continuous basis without interfering with the progress of the ship. In the days when he had lowered geophones to the bottom on the continental shelves, Doc had been lucky to get more than one shot per day. With the new technique, he could obtain dozens.

"The cruise had no elaborate plan," recalled Press. "We made it up as we saw what the bottom was. We worked up our data at sea in those days. We knew what we had. As soon as the seismic records were out of the photographic darkroom we had them analyzed—all day, all night, every day—and so called the next shot by the last one."

Doc was taking seismic shots as fast as humanly possible. David Ericson, a young paleontologist who had been the assistant of Harry Stetson, WHOI's sediment expert, described the process:

> I remember one used to hold the fuse in one's teeth, sitting on the afterdeck. The charge—a half pound of TNT—was in one hand and a lighter in the other. One wasn't supposed to put the fuse in until the last minute. Then you flung it. This continued at half-hour intervals day and night.
>
> [Ewing] would seize every opportunity to get information. There would be two-day intervals where he was going continuously. If the topography became surprising the ship would circle and try to outline it. With him you were either going to sink the ship with too much explosive or discover something interesting.

In order to prevent the TNT from sinking too deep before detonating, the charge was tied to a beer-can float—an empty beer can, of course, that had been soldered shut.

Such was the routine, twenty-four hours a day, seven days a week, for however many weeks they were at sea. Ewing expected that it would work that way, and he, being a hands-on leader demanding more from himself than from anyone else, made it work that way. Running a research ship was expensive, often costing tens of thousands of dollars per cruise, and he made sure that he got his money's worth of data every time. It was a hard life to be at sea with Ewing, but life was no easier on shore afterward. Once the ship reached port, the chief scientist, whether it be Ewing or someone else, packed the data in a suitcase and hand-carried it to wherever Ewing and his research team were based. The data were far too valuable to entrust to any courier.

PRIOR TO LEAVING Woods Hole on the Mid-Atlantic cruise, Ewing had borrowed a gravity corer built by Harry Stetson, WHOI's expert on sediments. Doc had offered to take a few cores and give them to Stetson when he returned. Along with the corer came Ericson. "I don't think anyone asked me to go," Ericson said. "I just stepped aboard and no one questioned me because it seemed the natural thing to do."

Ewing's work habits were far different from anything Ericson had experienced before at WHOI. "There was no English or Swedish social life in the main cabin that evening," Ericson said. You were working night and day. Every minute you could see [Ewing] was thinking of the data he was getting and the money he was spending."

Ericson, who later followed Ewing to Columbia, wasn't the only one having an epiphany on the cruise, however. Ewing had not previously been interested in deep-sea sediments. His offer to take cores was more or less a courtesy to Stetson. Doc thought the seamount east of Bermuda offered an excellent opportunity to attempt a core without wasting too much of the ship's time. He did not find the effort a waste.

"Well, some people, you know, never forget their first anything," Ewing later said, "their first drink, their first piece of tail. As it happened that core was one of the best of my life."

Ewing suddenly became addicted to obtaining cores—and was eventually to amass the largest collection of marine sediment cores in the world.

That first core was unforgettable. Ever since Bailey Willis's bombastic pronouncement of the permanence of the ocean basins, most geologists expected that—because the ocean basins had existed from the earth's birth, and because they were sheltered from the erosive processes that wore down the continents—a complete history of the earth would be found in the sediments of the ocean floor. The prevailing wisdom met its sudden demise as soon as Ericson examined the core. Rather than a complete record of the earth's past, Ericson found that the top eight inches consisted of relatively recent deposits. The rest of the core consisted of Eocene sediments laid down between 40 to 60 million years ago.

After leaving the area of the seamount, the *Atlantis* crossed over what is now known as the abyssal plain. For hours on end the depths recorded by the echo sounder hovered around 2,900 fathoms. Two and one-half days later, the sounder revealed the presence of low, scattered hills followed by higher rises divided by wide terraces. Ewing and company were over the foothills of the Mid-Atlantic Ridge. On August 1 Doc had two reasons to be excited: (1) *Atlantis* was over an area of jagged peaks and steep gorges; (2) he was notified by radio that his second wife, Margaret Kidder, whom he had married in 1944 in Woods Hole, had given birth to their third child.

The echo sounder on August 7 identified a sharp peak that rose two miles from the bottom of a deep gorge. Ewing, eager to get a sample of the top, sent down a corer. It was not retrieved until late that night. Unfortunately, it had not reached the bottom, so he ordered another one to be resent immediately. Eventually that one did make contact.

As the corer was being retrieved, Ewing noticed that one strand of the cable bringing it up had snapped and was beginning to unravel. It was a potentially dangerous situation—the *Challenger*'s first casualty on its famous cruise had been killed by a snapped cable—but Ewing figured the safest place to be was right at the break. He walked over and repaired the break, and the retrieval process continued. Near dawn the corer was recovered. The end was badly mangled from slamming into a hard surface, but a freshly broken rock was lodged within the damaged tip. Ewing now knew that few sediments obscured the ridge. He also realized that its origin must be volcanic, a fact confirmed by dredge samples later that day.

By the time *Atlantis* returned to Woods Hole on September 13, Ewing had accomplished a great deal. He and Press had refined his seismic technique so that the ship did not have to stop during a shot. Ewing had discovered that the Mid-Atlantic Ridge was topographically complex—composed of many sharp ridges and valleys—and, like mountain ranges on land, was bordered by low foothills and, farther out, flat plains. Doc had also proved that the ridge was made of volcanic rocks. Some of what he had found was quite puzzling, however. The sediments on the ocean floor were relatively young—and discontinuous in the time sequence represented—rather than continuously laid down since the earth's origin. What did the observations mean?

IN JUST TWO YEARS of residence at Columbia, Ewing and his rapidly enlarging group outgrew the pitifully small space available in Schermerhorn Hall. In addition to adding students and researchers, Doc hired Angelo Ludas, a veteran of the Manhattan Project, to run a machine shop. Ludas, with the help of Worzel, took excellent advantage of burgeoning government surplus lists to stock the shop with equipment and supplies. Ludas's mechanical skills enabled instrument designers like Ewing, Worzel, and Press, among many others, to fulfill their research visions for several decades after.

The lack of space for people and equipment, however, was straining Ewing's ability to achieve what he wanted. He was not the type to suffer limitations to his work for long. In 1948 the Massachusetts Institute of Technology offered him an estate near New Bedford, as well as the option to bring his entire staff and students, if he chose to join the faculty there and set up a geophysics program. Ewing seriously considered the offer, even taking Worzel, Press, Hamilton, and Steenland along on a visit to evaluate the New Bedford property.

About the same time, however, Florence Lamont, widow of financier Thomas Lamont, offered Columbia University their estate in Palisades, New York, about twenty miles north of the Morningside campus along the west bank of the Hudson River. Ewing, after returning from his visit to New Bedford, met with recently retired General of the Armies Dwight David Eisenhower, then president of Columbia, and Paul Kerr, chairman of the geology department. Eisenhower agreed to accept the gift of the Lamont estate—as long as it did not pose a financial burden to the university. Kerr agreed to raise $200,000 in outside funding to get a geological laboratory started. With financial arrangements taken care of, the estate was offered to Ewing.

Ewing's group debated the competing offers, eventually voting unanimously to stay at Columbia. The university received the deed to the property late in December 1948. Before long Press and Worzel appeared on the grounds to set up a seismometer in the estate's empty swimming pool. Press was especially excited in that the seismometer would no longer be rattled by vibrations from the subway line and traffic on Broadway, next to Columbia's main campus.

"We were all thrilled by the new space," Worzel later wrote. "The estate had 125 acres—actually more like 135 acres, but we had to give up ten acres that lay across the New Jersey state line. Robert Moses, New York's infamous road, bridge and parks builder, wanted the ten acres for his new Palisades Interstate Park system. Columbia wanted to close West 116th across its campus from Broadway to Amsterdam Avenue. They struck a deal."

Early in 1949 the rest of Ewing's group moved onto the grounds at the newly formed Lamont Geological Observatory. Tired of begging for sea time from other institutions, all Ewing needed now was his own ship.

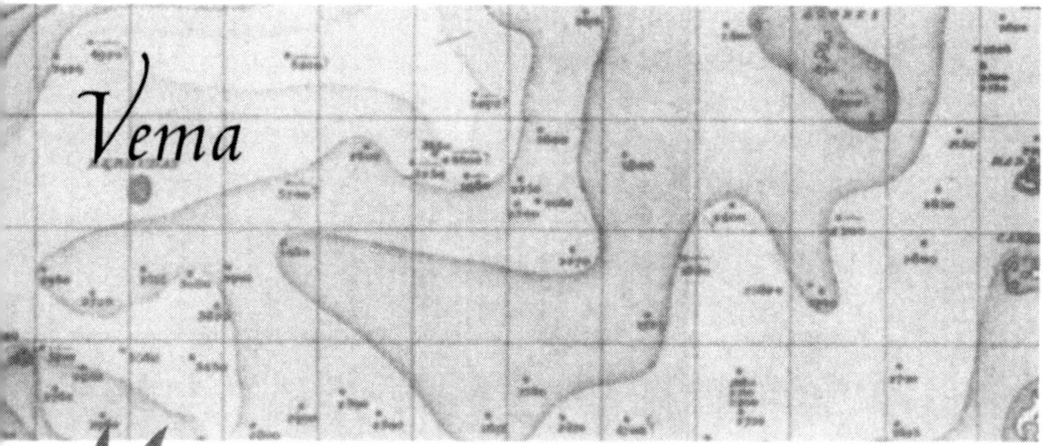

Vema

Maurice Ewing in 1951 convinced the Office of Naval Research (ONR)—which was established shortly after World War II to organize and conduct oceanographic research for the navy—that Lamont needed its own seagoing research ship. ONR agreed to provide an oceangoing tug, the *Allegheny,* to be shared by Lamont and Hudson Labs, which was also affiliated with Columbia University and was located across the Hudson River from Lamont. After Joe Worzel and Charles Lum "Chuck" Drake spent the rest of 1951 and part of 1952 supervising the conversion of the tug for oceanic research, and after Lamont graciously allowed Hudson Labs to take the ship on the first cruise in 1952, ONR told Ewing that Hudson Labs would need the ship the following summer for "classified" work. ONR later told Ewing that the *Allegheny* could not be spared, period.

Ewing made do in the summer of 1952, getting funds to charter another oceangoing tug, the *Kevin Moran,* for three months, then scrounging some time on the *Atlantis* later in the year. He could not afford to depend on other institutions for ships any longer, however. He and Worzel discussed the situation while traveling to and returning from a scientific meeting in Cambridge, England. They decided that Doc should tell ONR that since it had caused the problem, it should also pay Lamont the cost of chartering a substitute. Ewing dispatched Worzel to find the substitute.

Worzel bought a yachting magazine and browsed the ads for ships for sale or charter. He found one advertised by a Nova Scotia resident named Louis Kenedy. Worzel contacted Kenedy and decided to travel up to Lunenburg, Nova Scotia, to have a look at the ship. It was named *Vema.*

FINANCIER Edward F. Hutton needed a new yacht in the early 1920s and commissioned one from the Danish shipyard of Burmeister & Wain in Copenhagen. The ship, designed by Cox & Stevens of New York City, was to be a three-masted schooner, about 200 feet long, with a 33-foot beam, 17-foot draft, about 14,000 square feet of sail, and an auxiliary six-hundred-horsepower diesel engine capable of making eleven and one-half knots. The ship, *Hussar* (the fourth in a series), was launched in March 1923.

Hussar was luxurious, equipped with a Louis XV bedroom, an Edwardian sitting room with a marble-lined fireplace, a dining room with stained-glass windows, and gold faucets in the bathrooms. The ship, however, did not seem lavish enough seven years later when Hutton married Marjorie Merriwether Post, heiress of cereal tycoon C. W. Post. For her, he built a new sailing vessel: four masted, 320 feet long, and also christened *Hussar.* Hutton planned to sell the seven-year-old schooner.

One newlywed couple cast the schooner aside; another newlywed couple picked it up. Norwegian-born shipbuilding magnate Georg Unger Vetlesen and his wife, Maude, chartered the ship from Hutton and rechristened the schooner *Vema.* In May 1932, about a month after the Vetlesens were married, *Vema* smashed a twenty-seven-year-old record for the fastest crossing of the Atlantic—despite stormy weather. The former record holder, *Atlantic,* in 1905 sailed from Montauk Point to the Lizard, England, in twelve days and four hours. *Vema* arrived off Bishop's Rock ten days, twenty-one hours after passing Montauk Point.

One year later *Vema*, again with the Vetlesens on board, bested its own record by eleven hours.

Later in 1933 the Vetlesens purchased the schooner from E. F. Hutton, and they enjoyed the ship for eight years. By 1941, however, Georg Vetlesen's native country, Norway, had been brutally occupied by the Nazis. Vetlesen could see what was in store for the United States and could predict what his adopted home would need—ships. On February 15 the United States Maritime Service, a training agency operated by the Maritime Commission and the Coast Guard, announced that it had purchased the schooner from the Vetlesens. Two days later Vetlesen turned *Vema* over to the Maritime Commission; at that time the purchase price was disclosed: one dollar. The ship's estimated value at the time was about $100,000.

The Maritime Commission quickly ripped out the former palatial accommodations, creating bunk space for ninety-three trainees in addition to officers and crew. During World War II *Vema* served as a training ship for the Maritime Commission and an auxiliary patrol craft—a submarine hunter—off the United States's northeast coast. Auxiliary craft like the *Vema* weren't very good at hunting submarines—a consequence more of the military's not figuring out how to use the ships effectively than of any problem with the ships themselves. By the end of the war this former queen of the seas was relegated to the role of a floating bunkhouse for the Merchant Marine Academy at Kings Point, Long Island.

By the time Kenedy found the *Vema* stranded in a mud bank off Staten Island in 1952, the ship had been sold for scrap. He thought the bald-headed schooner (its topmasts had been removed) was too good to be slowly dismantled for pieces of metal here and there, so he bought it—for little more than its scrap value. With a four-man crew, and his wife cooking, Kenedy unfurled some of *Vema*'s sails and nursed the ship—whose engine was not working—to Nova Scotia.

"We had no heat aboard except a small oil-burning cook stove and the only machinery that worked was a small auxiliary engine which provided lights," Kenedy later recalled. "Fortunately we had generally moderate weather—though we had to run into Shelburne, Nova Scotia, and wait out a three-day snowstorm. The next westerly wind took us to the entrance of the Lahave River and we slipped over the bar at high water at three o'clock in the morning."

Kenedy's original intention had been to convert *Vema* into a freighting schooner, but, on his first thorough inspection of the ship at its new berth on the Lahave River, he changed his mind. In addition to the diesel engine, *Vema* was equipped with steam heat, electric winches, showers (hot and cold), and a radio. The galley included a freezer and room-size refrigerator. The schooner had enough fuel capacity to allow it to cruise for a month under power alone. Another nice touch—for Kenedy, at least—was the presence of a refrigerated water fountain in the mess room.

"When we got looking into the *Vema* we found that all the expensive equipment could be put into running condition with very little work," Kenedy said. "Though we'd saved the vessel from the boneyard, we soon realized that we'd be destroying a lot of her value if we stripped her down to a sailing freighter. So we refitted her as she was and began to shop around for a research organization that could use the *Vema* for an expedition."

WHEN WORZEL inspected the ship, he liked what he saw. *Vema* was bigger and had a lot more deck and cabin space than *Atlantis.* The engine was not working, but Kenedy assured Worzel that he would have it running in a week or two. Worzel asked Kenedy to put in a bid for a charter for two months starting and ending in New York. Kenedy said he could do it for twenty thousand dollars. Worzel returned to New York to obtain Ewing's and ONR's approval. Ewing liked what he heard and contacted ONR, who agreed to add an amount covering the charter price to Lamont's contract.

"I called Kennedy [*sic*] and asked him to come to New York to sign the necessary charter papers with the Columbia contracting group," Worzel later wrote. "He came down and we met at Lamont. While we were discussing the charter, I asked him what he would sell us the ship for. He answered $100,000 and that if we bought it before the charter was up, he would allow the charter fee to be deducted from the price. We concluded the charter and he returned to his home."

In April 1953 Kenedy, with a contingent of Lamont scientists aboard, set sail from New York, bound for the Caribbean and Gulf of Mexico. "We had an easy trip to San Juan, Puerto Rico, where we were joined by the *Atlantis* of the Woods Hole Oceanographic Institution," Kenedy said. "The two vessels were to work together, taking seismic readings of the ocean floor. To do this, one vessel would hove-to while the other sailed away on a certain course. Explosive charges were dropped every five minutes from the *Vema* and readings taken by the *Atlantis* up to a distance of forty miles. At that distance we were dumping over 300 pounds of TNT at a charge. On calm days, when the *Vema* was moving slowly under sail alone, we took a terrific pounding from the explosives. The *Vema* also took continuous depth recordings with a new-type fathometer accurate to within a fathom, or six feet, at a 3,000-fathom depth. And we towed a complicated device called a magnetometer to measure the magnetic characteristics of the ocean bottom."

For the most part, the crew and the scientists got along well—with one exception. Kenedy's mean-looking dog, a Labrador-husky mix named Gotlik—who could climb ladders as well as any of the men and who enjoyed making snacks of tropical birds that made the mistake of landing on "his" ship—had a unique way of making the scientists' lives difficult. Gotlik would slip into the head the scientists were using and eat their soap.

Engine trouble plagued *Vema* on the first cruise, however, and the diesel finally gave out in the western part of the Gulf of Mexico. The schooner continued on by sail until the research work was completed. One Sunday afternoon *Vema* called on the United States Navy base in Guantánamo Bay, Cuba, to load more explosives. A good breeze was blowing. With a large number of navy vessels at anchor, Kenedy decided it was showtime.

Vema entered the bay under full sail; then Kenedy made a ninety-degree turn and dropped all the sails—fore, main, mizzen, and three jibs—by himself and allowed the *Vema* to coast about a half mile upwind to the edge of a cluster of anchored ships. As the schooner began to lose way, he ordered the anchor to be lowered. The rope was let out slowly as *Vema* drifted downwind into perfect position in a line of ships—between two of the biggest. When Kenedy was through he got an ovation from sailors on the other ships.

The base itself was closed on Sunday, so Kenedy and Worzel had to wait until the following morning to make the required courtesy call at the commandant's office and to arrange for the delivery of the explosives. The base commander told the men that he had watched *Vema*'s arrival from his house and complimented Kenedy on his masterful ship handling.

Kenedy tried to get *Vema*'s diesel engine fixed while at the base. He quickly realized that the ship he had rescued from the mud off Staten Island would be an expensive one to maintain.

"The *Vema* had just too much machinery—and most of it required expensive shipyard work when it broke down," Kenedy said. "When we cracked the cylinder heads of the Diesel in the gulf, the United States Naval Base at Guantánamo, Cuba, tried to make repairs, but they didn't have the proper equipment. I finally had to order five new cylinder heads from Copenhagen—at $2,000 apiece. That sort of an operation was too rich for my blood. I wanted a ship I could maintain and, in general, repair myself."

Worzel, however, was pretty pleased with the ship, and he told Ewing so. He recommended that Lamont buy it.

BACK IN NEW YORK, Worzel supervised some refit and repair work on the ship while Ewing tried to obtain the funds for the purchase. On the last day of the charter, when the option to buy was to expire, Ewing called Worzel—who was with the *Vema* at Todd

Shipyard in Hoboken, New Jersey—and said he could not get the money.

Worzel at first accepted the news, but before long he decided it was time for action. He was unstoppable when he was riled. (One weekend at Woods Hole Worzel had used a fire ax to break through the door to the WHOI carpenter's shop and had "borrowed" some tools to make crates for shipping explosives to a ship scheduled to leave on an expedition from New London, Connecticut, that Sunday.) Worzel piled into his car and drove back to Lamont. Storming into Ewing's office, he informed Doc that if they did not get *Vema* right then, they might as well give up on ever getting a ship. Ewing turned to Frank Press, who was also present, and asked what he thought. Press agreed with Worzel.

Ewing, in desperation, decided to try to get Columbia University to put up the money to buy the ship—with the understanding that it would be repaid. He tried to contact university treasurer Joseph Campbell, but Campbell had already left for the day, and his secretary refused to tell Doc where Campbell had gone. But Ewing threatened her with dire consequences if she did not provide some information. She relented.

"Then I telephoned Cooperstown where he'd gone to rest, and got his wife," Ewing said. "He was playing golf and would be back at supper. That was too late. She said, 'If it's terribly, terribly important, I'll get him.' I said 'You'd better go,' and she went."

When Campbell called back, Ewing explained the situation. Campbell agreed to make arrangements to purchase *Vema* but called back a little after three that afternoon and said it was too late to do so through the treasurer's office. Instead—on the condition that Ewing say nothing about the deal until Campbell had a chance to talk to the university trustees—Campbell wrote a check of eighty thousand dollars from his personal account and bought the schooner.

Just when all appeared to be going well, however, Ewing got an angry call from one of the university trustees, who demanded to know why he had purchased a ship without the trustees' approval. Doc had to acknowledge buying *Vema* but asked how the angry trustee had learned of the transaction. Campbell, it turned out, had arranged for insurance to cover the ship once the transfer of ownership took place, and the order was placed through the trustee's firm. Campbell then smoothed everything over.

Doc had his ship. He almost, however, did not have much time to enjoy it.

ON JANUARY 13, 1954, the schooner—on its third cruise for Lamont—was caught in a fierce gale as it was sailing toward Bermuda to drop off some equipment for the Navy sofar station that Lamont operated. *Vema* pitched and rolled in mountainous seas as winds howled through the rigging.

Early that morning, Ewing was crossing the deck to the chart room when four fifty-five-gallon drums filled with lubricating oil broke free of their lashings and careened across the deck, damaging all in their path. Ewing, his brother John, First Mate Charles Wilkie, and Second Mate Michael Brown rushed to secure the barrels. Just as they thought the barrels would not break free again, a rogue wave slammed into the ship and washed the men and barrels overboard.

When Ewing came up, he saw Wilkie holding on to a barrel and his brother swimming toward a line. He tried to swim to a barrel himself but had swallowed much seawater and was too weak to make it. He began shedding his wet clothes before the weight dragged him under. As he took off his shoes, he wondered how long it would take them to reach the bottom three miles below and how they would look if later photographed by one of Lamont's deep-sea cameras.

Next Ewing heard Wilkie's voice, crying out "Doc! Help me," followed by a choking cough and silence as the doomed man drowned. The waves kept battering Ewing; each time he was hit he swallowed more water. As he breathed he heard the water bubbling in his lungs.

Vema was now about a mile away, but the ship was in good hands—Captain Don Gould of New Brunswick, New Jersey, and sailing master Fred McMurray, the former skipper of the *Atlantis*. Both had experienced their share of storms at sea and knew how to handle a ship in angry waters. Gould climbed to the crosstrees to spot the men in the water and McMurray—whom Ewing had lured from a New York City retirement home for this cruise—took the helm of the schooner, whose steering had failed for a time the day before. The schooner was rolling so badly that at one point Ewing saw one side of the bottom, from the keel up.

Gould and McMurray brought *Vema* back and stopped just upwind of John Ewing. No one saw him at first, despite his desperate yelling. The schooner was about to move on when someone spotted him. The crew threw him a line and pulled him aboard. The ship turned away from Doc, however. He doubted he would last much longer. Ewing couldn't swim—a blow to the neck when he was thrown overboard had left him partially paralyzed on his left side. He could not hold his breath and float—and he was fading in and out of consciousness.

"I guess you'd think that a person would be pretty much alone out there at a time like that," Ewing later wrote in a letter to his children. "I wasn't alone a bit. It seemed as though all the good people I love and who love me were there, and were encouraging me. Then they all went away and just you children were there, and it seemed that I needed to come and do something for my children. It seemed that all of you were about to drown, and I had to keep swimming to save you. Then only little Maggie was there. I couldn't see Maggie, but I could hear her. She was calling just the way she calls down the stairs when she hears my voice when I come home at night."

Someone else was calling, too. Brown was swimming toward Ewing, pushing along one of the fifty-five-gallon drums. He called out, "Doc, I could hold on to this barrel easier if you'd take hold of the other end." Ewing grabbed it and held on.

When the *Vema* next approached, someone tossed a line to Brown, who was quickly pulled to the side, dragging the drum and Ewing with him. At this point the schooner was rolling so heavily that the rails went under with each wave. Brown caught the rail on one of its downstrokes and rode it onto the deck.

As Brown went up, however, Ewing was pushed under. He saw a rope in the water beside him and grabbed it but could not pull himself up, because of his paralysis. On the next roll, the men on deck caught Ewing by the arms and hauled him to safety.

Vema's steering failed again as soon as Ewing was aboard.

John Ewing had an injured leg, which quickly healed. Doc was put ashore in a hospital in Bermuda and eventually recovered, but he had a slight limp for the rest of his life, which worsened when he was tired. Brown showered and stood his next watch.

Wilkie, however, was never found.

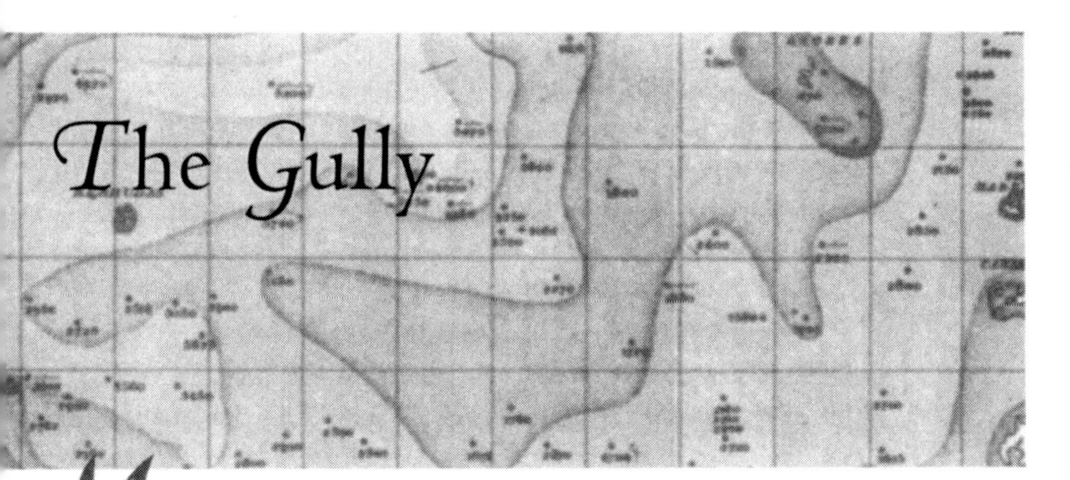

The Gully

Marie Tharp seemed rudderless as she searched for a good career while attending Ohio University in Athens in the early 1940s. Her father, William Edgar Tharp, told her to try to find something that she was good at but also, more important, that she enjoyed. She searched, and changed her major every semester, but nothing seemed to fit; nothing seemed interesting.

"I was looking for something that I could do well, and I was looking for something that someone would pay me to do," Tharp said. "And then I had to think about what women do, where can I get a job. In those days all we could do was to teach, be a secretary or a nurse. That's all a girl could get a job in.

"It didn't take me long to eliminate nursing because I couldn't stand the sight of blood. I didn't try to be a secretary because I couldn't type—I wasn't very good at that. So then I thought I was going to have to teach. . . . So I went around taking all these stupid courses in education."

Teaching seemed to be the only proper career for a young woman like her—or at least one of the few careers *open* to a young woman like her—until the full effects of the mobilization for World War II were felt at home. Millions of American men were in uniform fighting the Axis powers, leaving many industries at home depleted of manpower. Many of the jobs could not go unfilled, however, and women were invited to do "men's work." One of the professions hard hit by the war, yet highly important to the war effort, was geology. The industries that employed geologists felt the void, as did the institutions that trained them. Bodies were needed to fill the slots, and if that meant bringing women aboard, so be it!

One of the institutions that broke with tradition was the geology department at the University of Michigan, which sent out flyers promising jobs in the oil industry for women who completed its geology program. Tharp, eager for a chance *not* to teach, decided to take advantage of the school's fine offer. In 1943 she was one of ten women who entered the department. A little more than a year later she had earned a master's degree in geology as well as a job at Stanolind Oil and Gas Company (a forerunner of Amoco) in Tulsa, Oklahoma.

The work beat teaching, but Tharp was soon bored. She was not permitted to be much more than a clerk, compiling exploration reports from the field for others to discuss in meetings she was never invited to. Tharp began attending the University of Tulsa and earned a math degree. That was OK, too, but she sought something more—something to which she felt she could make a contribution.

By 1948 Tharp had had enough of Tulsa. She boarded a train for New York City, hoping to finally find an interesting job. Her first stop seemed a natural for someone with a degree in geology.

"I looked for work at the American Museum of Natural History," Tharp said, "but I decided I didn't want to work there after a paleontologist told me how it took two years to separate a fossil from the surrounding matrix. I couldn't imagine devoting so much time to something like that, so I tried Columbia [University] to see if I could get a more interesting research job."

The first people Tharp spoke with in Columbia's geology department were impressed by the fact that she had degrees in math and geology, and suggested that she talk to Maurice Ewing. Doc, however, was out at sea and would not be back for three weeks. Tharp went back to her apartment and waited for Ewing to return.

"When he heard about my background," Tharp said, "he was surprised and didn't know quite what to do with me. Finally he blurted out, 'Can you draft?'"

EARLY IN 1947 Ewing went on a Sigma Xi–sponsored lecture tour to speak on his work at sea. Doc's primary goal in going on the road was to find bright, well-to-do students to whom he could offer adventure instead of pay. One of his stops was at the University of Iowa. Bruce Charles Heezen, a junior interested in paleontology, attended Ewing's lecture. Later that night, over a tray of fossils, Heezen was introduced to Doc, who asked, "Young man, would you like to go on an expedition to the Mid-Atlantic Ridge? There are some mountains out there, and we don't know which way they run." Heezen heeded the call and abandoned a life studying fossils—or raising turkeys, like his father—for one on the sea.

In the summer of 1947 Heezen met Ewing at Woods Hole to begin preparing for a National Geographic Society–sponsored cruise of the *Atlantis* back to the Mid-Atlantic Ridge. After weeks of preparation, which included Heezen's involvement in building several new deep-sea cameras, Ewing informed the young Iowan that he would not be going on the *Atlantis*. Instead, he would depart on the *Balanus*, which had just become available. And Heezen—still one year away from graduation from the University of Iowa—would be going as chief scientist. His task: Take photographs of the continental margin off the east coast of North America.

Heezen was diligent: Despite repeated bouts with seasickness, he managed to obtain two hundred photos of the never-before-seen part of the abyss. He took the photographs back to Iowa to analyze the next year but ended up with more questions than answers. He would have to go to sea again. In the fall of 1948 Heezen entered Columbia as a student of Ewing's.

WHEN THARP BEGAN her drafting job at Columbia, she did work for whoever required her services. She was working exclusively for Heezen, however, in 1952 when he asked her to begin plotting depth profiles of the North Atlantic. After compiling soundings from several research cruises, Heezen hoped to have enough data to determine, at least in part, which way the mountains of the Mid-Atlantic Ridge ran. He and Tharp had plenty of data for the initial task. As late as the 1930s a collection of a few hundred soundings on a cruise was thought to be enormous. But with technological advances, and Ewing's drive and direction, tens of thousands of depth measurements in the North Atlantic had been acquired from 1946 to 1952. Most of the data were obtained on cruises of the *Atlantis.* The 1922 cruise of the navy destroyer *Stewart* was another important source of sounding data.

Tharp and her research assistant, Hester Haring, went to work at drafting tables in a room on the second floor of Lamont Hall, the former residence of Thomas and Florence Lamont, which was now the main building of Ewing's fledgling geological observatory. First, Haring would replot the sounding data using the charts compiled by the ship's officers and scientists. Then she and Tharp plotted profiles of the topography along the ship's course. The profiles had to be drawn in a consistent manner. The horizontal scale was forty miles to the inch. The depths were plotted at a vertical scale of one inch to one thousand fathoms. The west end of the profile had to be on the left, the east end on the right. Of course, the depth values had to be correct, too. Any mistakes and someone like Tharp or Heezen would angrily scrawl a message like "Plotted Backwards!" on the profile and have it redrawn. Tharp or Heezen would then compare the depths on the checked and corrected profiles with the original sounding records.

Eventually, after the drawing, checking, correcting, redrawing, rechecking, and so on, was done, Tharp was left with a hodge-podge of disjointed, if not disconnected, profiles of sections of the North Atlantic floor. Plotted on a map, the ships' tracks looked like a web woven by a disturbed spider—with the eccentricities in the paths often caused by ships' sometimes unsuccessful attempts to dodge storms. The web's rays radiated out from Bermuda, where most of the research vessels took on supplies and water. Heezen asked Tharp to order the chaos and make several more or less parallel profiles of the North Atlantic.

Tharp, faced with an almost impossible knot of ships' tracks to untangle, took copies of the profiles and cut them into sections. Next she arranged the sections in proper order from north to south and west to east. Her task became even more difficult where two tracks crossed one another, such as, for example, when a southeastward track cut across a northeastward one over a seamount. Tharp would take the section of the first track northwest of the seamount and the section of the second track northeast of the seamount and recombine them into a new transect that lay entirely north of the undersea feature. She would repeat the process with the southwest limb of the second track and the southeast limb of the first track so that the resulting recombined transect, more or less parallel to the first, lay entirely south of the seamount. In time, Tharp carefully reduced the jumble to six west-to-east transects across the ocean.

As Tharp plotted the more northern profiles, she noticed a large valley at the center of the Mid-Atlantic Ridge. Although the valley wasn't as prominent in the three southern profiles, it was still there, typically one thousand fathoms deep and nine to thirty miles wide—as deep as the Grand Canyon, but much wider. In the six weeks it took her to prepare the profiles, Tharp became convinced that she was looking at a rift valley and told Heezen so.

Heezen did not want to hear the news. A rift valley indicated that the earth's crust was spreading apart, and that might mean the continents on either side of the Atlantic were getting farther and farther away from each other. If the continents were drifting apart, one would have to conclude that there was something to Alfred Wegener's crackpot theory. Speaking out in favor of continental drift would be an act of professional suicide.

Heezen looked at Tharp's profiles. No matter how hard he tried to make the rift valley disappear, it would not. He groaned, and said, "It can't be. It looks too much like continental drift."

Although Heezen was skeptical of Tharp's interpretation of the data, she knew what she saw and was determined to bring him around to her way of thinking.

AT ABOUT THE SAME time Bell Laboratories asked Heezen to map the locations of breaks in transatlantic telephone and telegraph cables. Bell Labs was planning to lay more cables and wanted to know if midocean earthquakes were responsible for the damage. Heezen hired Howard Foster, a recent graduate of the Boston School of Fine Arts, to plot the locations of recorded earthquakes around the world. Another group was put to work plotting the locations of cable breaks.

Heezen was adamant that all oceanographic data be plotted at the same scale. His instructions paid off when Tharp, working at a drafting table next to Foster, noticed that the earthquakes Foster had plotted were roughly in the same region as her rift valley. There were some differences, but the earthquake data were subject to error because of the limited seismic network used in recording the locations of the epicenters. Heezen wasn't impressed with the matchup at first, but, when he took into account the extent of the probable error in locating the earthquakes, he realized that the location of Tharp's rift valley coincided with the location of seismic activity—which coincided with the cable breaks.

By that time, almost a year after Tharp had pointed out the rift valley, Heezen no longer dismissed her interpretation of the depth profiles as "girl talk." With the three types of data they were plotting all in agreement, the conclusion was obvious.

Tharp was right.

Ewing also began to get interested at this point. He had heard of this "gully," as they called it, and he would pop into their lab from time to time and ask, "How's the gully coming?" It was coming along just fine. Using sounding data from the *Meteor* expedition, Tharp had extended her profiles of the Mid-Atlantic Ridge and rift valley into the South Atlantic. Data from other expeditions revealed similar features in the Indian Ocean, Arabian Sea, Red Sea, and Gulf of Aden. A United States Navy expedition had found a large north-south ridge system in the eastern Pacific.

While Tharp and Haring busied themselves with sounding data, Foster was plotting tens of thousands of earthquakes around the world. The pattern held. Wherever there was a midoceanic ridge, there were earthquakes. When the Indian Ocean earthquake belt was shown to be continuous with the East African Rift Valley, there was but one conclusion.

In 1956 Ewing and Heezen reported the results of the work at a meeting of the American Geophysical Union. The title of the paper was "The Mid-Atlantic Ridge Seismic Belt." Only the abstract from the paper was published. Small as the abstract was, and as economical as the language was, it was to have a tremendous effect on the earth science establishment.

> The Atlantic belt of earthquake epicenters follows the crest of the Mid-Atlantic Ridge and its prolongations into the Arctic and Indian Oceans with a precision which becomes more apparent with the improvement of our knowledge of the topography and of epicenter locations. These are all shallow shocks. Their apparent departure from the narrow crest of the ridge seldom exceeds the probable error of location. The crest is 30 to 60 miles wide, very rough, and on a typical section shows several peaks at depths of about 800 to 1100 fathoms. There is usually also a conspicuous median depression reaching depths of about 2300 fathoms. This is interpreted as an active oceanic rift zone which continues through the African rift valleys.

PRINCETON GEOLOGY students and faculty gathered in an auditorium in Guyot Hall on the night of March 26, 1957, to hear Bruce Heezen talk about the rift valleys that he and Tharp had discovered. Heezen wasn't one to waste time preparing for a talk: His discussion was as disorganized as he was disheveled. Students nodded off here and there, daydreamed, or wondered why the faculty members thought he was so important—or why the talk was so important.

Harry Hess, who in 1950 had become chairman of the Princeton geology department, paid attention, however. He realized that Heezen was onto something huge, something revolutionary—more so, probably, than Heezen himself.

At the end of the talk, Hess stood up and said something that shocked the students to attention. Hess looked at Heezen and said, "Young man, you have shaken the foundations of geology!"

Shaken—and Stirred

Harry Hess was ready to change his preconceived notions to fit the facts. The challenge, however, was to determine what they meant. Thanks to the boom in oceanographic research in the years following World War II, Hess had a lot of facts to consider—most of them coming from Maurice Ewing and his associates at Lamont.

Ewing drove Lamont's scientists and ship hard. One time, when British geophysicist Teddy Bullard asked Doc where he kept his ships, Ewing answered, "I keep my ships at sea." Ewing did not think *Vema*—and later the *Robert D. Conrad*—should spend more than thirty days a year in port. "You don't collect much data when your ship was in port, tied to the dock," he once told a WHOI colleague who made the mistake of poking fun at Lamont's lack of a proper home port. Ewing also found ways to get the maximum amount of data possible. While one of his ships was underway, Doc wanted the continuous echo sounder—or its vastly improved replacement, the precision depth recorder developed at Lamont by Bernard Luskin—running continually. Such was also the case with the seismic profiling equipment, a gravimeter and a towed magnetometer. From one end of the stern, the seismic hydrophone string would be towed near the surface behind the ship, while the magnetometer would likewise be towed near the surface from the other side. As the ship slowed to stop at a station, scientists would be on deck getting their equipment ready to be deployed. When the ship stopped, lines would immediately be dropped from both sides of the deck—which, on the *Vema*, were only about thirty-five to forty feet apart on the surface—to the ocean floor thousands of fathoms below to obtain sediment cores, heat flow measurements, ocean bottom photographs, or some other type of sample or information. Not a second was wasted.

Lamont and Woods Hole were not the only major oceanographic institutions involved in studies of marine geology and geophysics. Scripps Institution of Oceanography in La Jolla, California, the oldest of the three, began a major effort to expand its focus from coastal waters to the deep oceans after World War II. Roger Randall Dougan Revelle, who became Scripps's fifth director in 1951, oversaw an expansion of the institution's fleet and coaxed its scientists beyond the continental shelf. Henry William Menard almost single-handedly revolutionized studies of the Pacific Ocean floor. Like Bruce Heezen and Marie Tharp at Lamont, he was mapping and discovering features never before seen or imagined, including massive fractures in the Pacific Ocean floor where large blocks of oceanic crust slid past each other. The first fracture zone to be documented by Menard, the Mendocino Fracture Zone, originates near the coast off Cape Mendocino, California, and extends for hundreds of miles to the west. Its most noticeable feature is a sheer, six-thousand-foot high escarpment on the ocean floor.

Arthur E. Maxwell, who had worked with Teddy Bullard in developing a deep-sea heat probe after World War II, began collecting hundreds of heat flow measurements on Scripps's expeditions. Russell Watson Raitt obtained numerous seismic profiles, and Ronald George Mason, of Imperial College in London, and Arthur Datus Raff, of Scripps, began magnetic surveys on Scripps's ships in the North Pacific.

The increase in available data was phenomenal.

"By 1956 Lamont, using Navy submarines, had tripled the number of gravity observations at sea," Menard later wrote of the postwar explosion in data collection. "Scripps had taken none. There were no heat-flow observations in the oceans in 1946, but by 1963 Scripps had taken about 300 and Lamont had taken none. There were perhaps 100 cores of sediment from deep-ocean basins in 1948. By 1956 Lamont had taken 1,195. Ewing was obsessed with cores and Lamont always led the world, but by 1962 Scripps had about 1,000. There had been no seismic stations in the deep sea, and by 1965 there were hundreds. Underway data, essentially continuous observations, had multiplied even more. The number of deep-sea soundings had increased by about 10^8 and the number of plotted soundings by 10^5. Nothing comparable to shipboard magnetic profiles had ever been known, and Lamont and Scripps had towed magnetometers for hundreds of thousands of miles."

BY NOW virtually everyone recognized that a rift valley was a tensional feature—an area where the earth's crust was splitting apart—and that magma welled up from the depths into the crack to form new crust. That a rift zone lay at the heart of the vast submarine mountain range encircling the earth meant that an as-yet-unheard-of amount of crust was being formed worldwide.

The last stubborn vestiges of contraction theory were finally eradicated by the discovery of a world-encircling rift zone—how could the earth be shrinking amid the creation of all this new crust? Permanence theory (the idea that ocean basins never changed), itself terminally ill, exited without even a whimper. The ultimate question at this point was to figure out what, if anything, was happening to the old crust.

After the discovery of the midocean rift, Bruce Heezen became the first at Lamont to officially break with the concept of the permanence of ocean basins. He began advocating an expanding earth—which makes perfect sense if new crust is being created while none is being destroyed. Under those conditions, expansion is the only possible outcome.

Heezen wasn't the first to embrace the concept of an expanding earth. A German, Ott Cristoph Hilgenberg, in a 1933 monograph, *Vom wachsenden Erdball* (The expanding earth), proposed that the breaking apart and drifting of continents could be explained by expansion. J.K.E. Halm, a South African astronomer, followed up Hilgenberg by looking to the stars for guidance. In an address before the Astronomical Society of South Africa, published in 1935, Halm pointed out that extremely high densities of matter can be achieved when atoms are stripped of their electron shells—a phenomenon that occurs in some types of stars—and suggested that as a result of a similar process the early earth was much denser than it is now and the mean diameter much smaller: 6,750 miles rather than about 7,900 miles today (the difference in circumference would be 21,000 miles during a period of higher density rather than 25,000 miles at present).

A Nobel prize–winning physicist presented an argument that made expansion theory look more probable. In 1938 English-born Paul Adrien Maurice Dirac proposed, in an article titled "A New Basis for Cosmology" in the *Proceedings of the Royal Society*, that the gravitational constant would decrease as the age of the universe increased. Robert Henry Dicke, a physicist at Princeton University, while exploring alternatives to Albert Einstein's theory of general relativity, proposed in a paper in *Science* in 1959 that the earth would expand somewhat in response to a decrease in the gravitational constant. On another occasion he suggested that oceanographers seek cracks in the ocean floor that would indicate that such an expansion had taken place.

Expansion theory never quite caught on, however, until the 1950s, when Laszlo Egyed, director of the Geophysical Institute in Budapest, Hungary, and Samuel Warren Carey, a geologist at the University of Tasmania, came to its defense. Egyed in 1956 published a paper, titled "The Change of the Earth's Dimensions Determined from Paleogeographical Data," in which he suggested that the earth's core is in an unstable, high-pressure state. As the core degenerates into a low-pressure phase, the density decreases; thus the volume increases. The magnitude of expansion he envisioned exceeded by several orders of magnitude anything proposed by Dicke. Carey, organizer of a symposium on continental drift in Tasmania in 1956, concluded in his contribution to the event that only an expanding earth could explain the change in the distribution of landmasses from the Paleozoic era to the present. Carey, armed with an extensive knowledge of global geology, became a formidable proponent of expansion theory.

To Heezen, expansion had a number of benefits: It accommodated the crust created at the ridges; it could account for "polar wandering" patterns documented by paleomagnetism researchers in England, such as Nobel laureate Patrick Maynard Stuart Blackett (later Baron Blackett of Chelsea) and his protégé, Stanley Keith Runcorn; and it could explain continental drift without requiring the continents to plow through the ocean floor—as Wegener had proposed—but allowing them instead to float passively along with the "drift" of the basement below.

While expansion theory took care of a number of problems, Hess was not eager to accept it.

WHAT IF the earth wasn't expanding? Then crust must be destroyed somewhere. Harry Hess was in a particularly good position to guess where. Hess had worked with Felix Andries Vening Meinesz on gravity surveys of the deeps adjacent to island arcs—and well knew the negative gravity anomalies found in conjunction with the trenches. He was an early supporter of Vening Meinesz's tectogene concept, where the crust of the earth buckles downward in response to descending limbs of convection currents.

Hess began to weigh the abundant new evidence. Menard had earlier suggested that midocean ridges tend to be found in the middle of the oceans. Bullard, Maxwell, Revelle, and Richard Von Herzen had reported in the late 1950s that heat flow through the ocean floor was much greater than what could be expected by conduction from the mantle alone, and it was highest at the midocean ridges. Only two possibilities could explain the phenomenon. One was that the rocks of the ocean floor were much more highly radioactive than the rocks of the continents—but by then it was pretty obvious that the ocean floors were made primarily of basalt, which contains very little radioactive material. Thus, by default, the prime suspect for the observed heat flow was—convection currents!

Doc Ewing, along with his brother, John, had reported from their seismic studies that the sediment cover on the Mid-Atlantic Ridge was exceedingly thin. Raitt found a similar pattern on the East Pacific Rise. They had also found that sound velocities were much lower than expected on the ridge crest. Hess had previously asserted that the lower velocities along the ridge crest were due to a change in the density of mantle material from higher density olivine to lower density serpentine, following a reaction with water called serpentinization. The less dense serpentine would rise above the surrounding seafloor. Serpentinization is a reversible reaction: In time serpentine would revert to olivine, and the rock would contract and sink until achieving isostatic equilibrium. Hess discarded the serpentinization idea following a talk with another Princeton geologist, Carl Bowin, whose experiments indicated that the hypothesis was incompatible with what was then known about the structure of the crust and mantle.

As Hess quietly—he did everything quietly—considered what Bowin had told him as well as the heat flow data, he concluded that the structure of the midocean ridges could not be explained by chemical processes. A thermal process could, however. As the hot magma welled up from the depths, it would cool on contact with the water. Still, the high heat content in the newly solidified rock would make it relatively less dense that the surrounding crust. It would continue to rise until achieving an isostatic equilibrium. In time, however, the rock would cool further and become more dense, sinking to eventually form the basement of the abyssal plains.

Hess, in a flash of insight, put all the disparate bits of evidence together into a coherent theory.

"The unexpected regularities in the spherical harmonics of . . . the earth's topography might be attributed to a dynamic situation in the present earth whereby the continents move to positions dictated by a fairly regular system of convection cells in the mantle," he wrote in his manuscript "History of Ocean Basins," which was finally published in 1962 after existing in preprint form for nearly two years. Hess then explained how convection currents explained Menard's observation that submarine ridges tend to occur in the middle of ocean basins. "The mid-ocean ridges could represent the traces of the rising limbs of convection cells, while the circum-Pacific belt of deformation and volcanism represents descending limbs. The Mid-Atlantic Ridge is median because the continental areas on each side of it have moved away from it at the same rate—1 cm/yr. This is not the same as continental drift. The continents do not plow through oceanic crust impelled by unknown forces; rather they ride passively on mantle material as it comes to the surface at the crest of the ridge and then moves laterally away from it. On this basis the crest of the ridge should have only recent sediments on it, and recent and Tertiary sediments on its flanks; the whole Atlantic Ocean and possibly all of the oceans should have little sediment older than the Mesozoic."

Hess offered one test of his hypothesis. During World War II he had discovered the flat-topped mountains called guyots. Near the midocean ridges, the tops of the guyots are near the surface, but as the distance from the midocean ridge increases, the tops are found at deeper and deeper levels until they reach the edge of trenches, where they disappear. Hess once attributed the phenomenon to serpentinization and its reverse reaction, but in "History of Ocean Basins" he suggested that guyots are in fact volcanoes that used to be higher—that in fact broke the surface of the water along the ridge crests. Their flat tops are due to erosion at the surface of the water. As the crust moved away from the rifts, cooled, and subsided, the volcanoes became inactive and sank. Since guyots were found only between a ridge and a trench, but not among island arcs on the opposite side of a trench, Hess suggested that the mountains were carried down as the crust they lay upon descended into the trenches by subduction.

Hess's ideas were radical, and he knew it. Rather than risk offending his colleagues by coming out swinging, as Alfred Wegener had done, Hess defused most hostile reactions to "History of Ocean Basins" with a masterful introduction:

> The birth of the oceans is a matter of conjecture, the subsequent history is obscure, and the present structure is just beginning to be understood. Fascinating speculation on these subjects has been plentiful, but not much of it predating the last decade holds water. Little of [Johannes Hermann Frederik] Umbgrove's (1947) brilliant summary remains pertinent when confronted by the relatively small but crucial amount of factual information collected in the intervening years. Like Umbgrove, I shall consider this paper an essay in geopoetry. In order not to travel any further into the realm of fantasy than is absolutely necessary I shall hold as closely as possible to a uniformitarian approach; even so, at least one great catastrophe will be required early in the earth's history.

The "catastrophe" is an overturning of the contents of the early molten earth that caused the formation of the core and protocontinents. Umbgrove's "summary" is a 359-page monograph, *The Pulse of the Earth.*

HESS'S HYPOTHESIS was widely circulated as a preprint and was presented as a report to the Office of Naval Research in 1960. It was supposed to be published in a mulivolume opus, titled *The Sea,* but Hess withdrew it as a result of repeated publication delays. He then submitted it to a volume to be published by the Geological Society of America honoring fellow Princeton geologist Arthur Buddington, which is where it finally appeared in 1962.

In the meantime Robert Sinclair Dietz, who had studied at Scripps before World War II and who was at the time a researcher at the Navy Electronics Laboratory in San Diego, published similar ideas in a short paper, "Continent and Ocean Basin Evolution by Spreading of the Sea Floor," in *Nature* in 1961—after Hess's preprint had been widely circulated but before it appeared in the Buddington volume. That led to two controversies. One was whether Hess or Dietz should have priority for the hypothesis. Dietz later defused the matter in 1962 by writing that Hess should be regarded as the originator. Another controversy concerned what Dietz knew of Hess's work and when he knew it. Menard, who was asked to comment on both manuscripts, later maintained that he told Dietz of Hess's work after Dietz had already written a draft of the seafloor-spreading paper. Hess later wrote that Dietz had gotten the idea from informal discussions they had on the matter in 1960. Dietz, while ceding priority to Hess, does not appear to give Hess any credit for influencing his thoughts—he in fact makes no mention of Hess's work on anything other than guyots in his 1961 *Nature* paper and in a subsequent contribution to a symposium on continental drift edited by Runcorn and published in 1962. When he did confer credit to Hess for coming up with the hypothesis—in another symposium volume published in 1962—he did so in a "Note added in proof" at the urging, according to Menard anyway, of Menard and others.

Despite the controversies, Dietz made important contributions to the subsequent debate. First, he coined the term *seafloor spreading,* which the hypothesis is known by. He also elaborated on terrestrial analogues to spreading seafloors: Continental crust sitting over converging convection currents in the mantle will buckle, giving rise to mountain belts. Over diverging currents, continents will rift apart. In converging areas where less dense continental crust rides up over denser oceanic crust, or sima—as on the West Coast of North America—mountain belts will form, but if the continental block rides passively atop a mantle current along with the sima, the margin will be tectonically stable—as it is along the East Coast of North America.

Ironically, Wegener had foreseen that convection currents could serve as a mechanism for drift, but he never had enough data to justify doing anything more than give convection a mention. Arthur Holmes, however, had done more, showing how convection could conceivably serve as a driving force for drift. But his ideas on the topic, which had most recently appeared in a textbook he wrote in 1944, had been almost completely forgotten in North America by the time Hess and Dietz were hard at work on their hypotheses. Some, like Arthur Augustus Meyerhoff, have since argued that Holmes deserves priority for the concept—although Meyerhoff seemed more motivated by disdain for anything related to continental drift rather than concern for Holmes's reputation.

IN 1962 Harry Hess spoke on seafloor spreading at a scientific meeting in Cambridge, England. A young student there, Frederick John Vine, was intrigued by the presentation. He thought to himself that there must be a more convincing way to test Hess's hypothesis.

The Paleomagicians

The magical powers of the mysterious magnetic rock known as lodestone had puzzled the natural philosophers of the East and West for millennia. Even after these powers were harnessed—in a device known as the compass—to save the lives of untold numbers of voyagers at sea and on land, few understood how lodestone attracted some rocks and repelled others or why a compass needle generally pointed north.

In 1269 Peter Peregrinus, a member of the army of the King of Sicily, Charles of Anjou, had been researching the properties of lodestones while the army besieged the city of Lucera in what is now southeastern Italy. Through a series of ingenious experiments with floating lodestones, he demonstrated that they had two "ends"—and coined the term "polus," or pole, to differentiate between the two—and that floating lodestones tend to orient themselves in a north-south direction. He showed how lodestones attracted and repelled each other: if the facing ends of two lodestones are of opposite polarity (north-south), they will come together, but if they are of the same polarity (north-north or south-south), they will push themselves apart. Peregrinus also demonstrated how a lodestone could be used to magnetize a needle of iron.

In time, mariners and other travelers had discovered that compass needles do not point toward true geographic north but to a magnetic pole on an axis slightly askew of the earth's rotational axis—and the degree of offset from true north varies from place to place. In 1576 Richard Norman, an English instrument maker, found that a floating compass needle tends to point slightly downward.

William Gilbert, a London physician whose hobby was the study of lodestones, began dispelling some of the mysteries of magnetism in the latter portion of the 1500s. He devoted many years of study to magnetic phenomena. He discovered that lodestones attracted iron with one end and repelled it with another, that iron lost its magnetic abilities when heated to a red-hot state, but that wrought iron could become magnetized by being repeatedly struck with a hammer. He also concluded that the earth itself was magnetized and that the magnetic powers of lodestones were somehow related to those of the earth itself. Gilbert published his findings in a treatise, *De Magnete,* in 1600.

In 1691 Anglo-Irish chemist Robert Boyle discovered that rocks or other solid materials that have cooled from an extremely heated state may preserve a record of the orientation of the earth's magnetic field. (The temperature at which the magnetic orientation is locked into the material is known as the Curie point, for Pierre Curie—codiscoverer of radium along with his wife, Marie Curie.) The phenomenon, called "remanent," or residual, magnetism, can be used to determine the position of the earth's magnetic pole at some point in the past. In 1853 Macedonio Melloni studied lava flows of the Phlegraean Fields and Mount Vesuvius in Italy and found that remanent magnetism in the rocks roughly paralleled the earth's magnetic field. Melloni heated volcanic rocks to red hot and then allowed them to cool. Even though inserted in the ovens at random orientations, upon cooling their magnetic fields were again aligned parallel with that of the earth. Giuseppe Folgheraiter in 1899 confirmed Melloni's conclusion, this time testing brick and Etruscan, Greek, and Roman pottery.

The study of remanent magnetism took a more interesting turn in 1909 when Bernard Brunhes discovered that some lava flows in the Massif Central in France have a magnetic polarity exactly opposite to, or *reversed* from, that at present. Raymond Chevallier in 1925 carefully analyzed the magnetic patterns in dated lava flows on Mount Etna in Italy, comparing them with twelfth- to seventeenth-century magnetic records from nearby observatories. At about the same time Paul-Louis Mercanton suggested that if the normal and reverse magnetization patterns found in rocks were caused by flip-flops in the polarization of the earth's magnetic field, then the reversals should be found at identical periods worldwide. (Thus, during a period of reversed magnetization, a compass needle would point south rather than north.) Mercanton was the first to appreciate how paleomagnetic studies could contribute to questions of polar wandering and continental drift.

Most of the work on paleomagnetism at the time was being conducted in Europe. In 1929, however, Motonori Matuyama of Kyoto Imperial University, Japan, reported the discovery of reversed magnetism in 700,000-year-old rocks from Japan and Manchuria. He determined that the switch to reverse magnetization occurred in the early Pleistocene, and that the polarity switched back to normal afterward. Matuyama's work was especially important because he used the age of the rock layers to determine the timing of the magnetic reversals. By following Matuyama's example in more and more regions, a chronology of periods of normal and reverse magnetization could be compiled.

Teddy Bullard, with German-American physicist Walter Maurice Elsasser, helped develop the dynamo theory, which describes how the earth's magnetic field is generated by the motion of molten iron in the earth's outer core. Patrick Blackett developed a competing theory to Bullard and Elsasser's dynamo in which the rotation of a body—like the earth—is alone sufficient to generate a magnetic field. To test the idea, Blackett had to devise an extremely sensitive magnetometer to measure the strength of a magnetic field. He also had to borrow nearly forty pounds of pure gold from the Bank of England—try getting away with that nowadays!—with which to make an object of theoretically sufficient mass to generate a field to measure. Blackett discovered no field, thus dooming his theory. The Bullard and Elsasser dynamo prevails, but Blackett's new magnetometer, called an astatic magnetometer, allowed him to study magnetism in samples of rocks.

Blackett, as he began to study magnetism in rocks, soon realized that the patterns he was seeing might hold the key to answering the question of whether continental drift, in some form, had actually occurred. In a talk in Jerusalem in 1954, Blackett predicted, "Major countries will have to study the magnetism of their own rocks just as they do their own geology. I have no doubt at all that the result of their work will, in the next decade, effectively settle the main facts of land movements, and in so doing will have a profound effect on geophysical studies of the Earth's crust."

Blackett dispatched his students around the world to collect samples and bring them back for analysis. If detailed notes had been made of the position of the sample in the field before it was extracted from the surrounding rock, his astatic magnetometer could determine the three-dimensional orientation of the magnetic field at the time the rock cooled.

By the time of Blackett's lecture in Jerusalem in 1954, three members of his research group, John A. Clegg, Mary Almond, and Peter H. S. Stubbs, had found something interesting: The magnetic needle in 200-million-year-old rocks from England pointed in a different direction than in rocks of the same age from other regions. They concluded that England had rotated thirty-four degrees clockwise since then. Other evidence indicated that Spain had rotated away from the west coast of France, thus opening up the Bay of Biscay.

Keith Runcorn did his fieldwork in the United States, particularly studying lava flows in Oregon and the shale beds of the Grand Canyon. In rocks such as shales, fine-grained materials rich in magnetic elements such as iron will settle in an orientation determined by the prevailing magnetic field. Runcorn used this and other data to reconstruct the migration of the North Magnetic Pole. He also deduced that the earth's spin axis, too, had migrated as a result of a disruption to the earth's equilibrium by mountain building or convection currents. Before long, however, he concluded that the continents themselves had done the moving. By 1960 he was a firm believer in—and one of the foremost advocates of—continental drift.

PALEOMAGNETISM REMAINED such a poorly understood phenomenon that paleomagnetists were referred to by some as paleomagicians. At the time that Blackett and Runcorn embraced continental drift, a controversy raged over whether magnetic reversals were caused by a flip in the polarity of the earth's magnetic field (field reversals), or whether the reversals were caused by local or mineral-specific factors (self-reversals). The controversy did not affect the research of the Blackett and Runcorn group, however, as they were concerned only with the direction of the magnetic field, not the polarity.

Allan Verne Cox and Richard Rayman Doell began studying paleomagnetism at the University of California at Berkeley. By the summer of 1958 Cox, who had been working part time for the United States Geological Survey in Menlo Park, California, and Doell, who had left a position at the Massachusetts Institute of Technology and returned to California following treatment for cancer, began to discuss plans for setting up a magnetics lab at Menlo Park. They had been in touch with James Robinson Balsley, Jr., chief of the Geophysics Branch of the survey, who was interested in learning whether or not magnetic reversals were related to changes in the polarity of the earth's magnetic field. Balsley was a pioneer in magnetic studies, having developed an airborne magnetic detector for use in geophysical studies after World War II. Balsley gave his blessing to Cox and Doell, and the Rock Magnetics Project was born.

Shortly thereafter Cox and Doell brought G. Brent Dalrymple aboard to help with radiometric dating of rock specimens. Their primary goal was to resolve the field- versus self-reversal controversy. In order to do so, however, the three had to develop a magnetic reversal timescale.

They had competition. Another research group, headed by Ian McDougall and Donald Harvey Tarling, formed shortly afterward at the Australian National University in Canberra. McDougall and Tarling were likewise working to develop a magnetic reversal timescale. A race was on.

Cox, Doell, and Dalrymple won the first leg. On June 15, 1963, their article, "Geomagnetic Polarity Epochs and Pleistocene Geochronometry," appeared in *Nature*. In it they reported that self-reversals were unlikely. They also presented the first, but very crude, magnetic reversal timescale, which left one crucial question unanswered—whether the duration of periods of normal or reversed polarity were regular or irregular. If the earth's magnetic field flip-flopped on a regular schedule, a magnetic timescale would be difficult to use for dating purposes. If the duration of the reversals was irregular, however, a magnetic timescale would be invaluable. As when a tree-ring scientist dates wood by matching the pattern of growth rings in a sample to the growth variations documented in a previously developed tree-ring chronology, or when a forensic scientist identifies a suspect by matching the ridge patterns in fingerprints taken from a crime scene to those of fingerprints stored in a police database, a paleomagnetics researcher could theoretically date a series of reversals by matching their pattern against the known magnetic timescale.

McDougall and Tarling responded quickly. Their article, "Dating of Polarity Zones in the Hawaiian Islands," appeared in the October 5, 1963, issue of *Nature*. With samples from basalt outcrops in the Hawaiian Islands, they demonstrated that the pattern of magnetic reversals was more complicated than that indicated by Cox, Doell, and Dalrymple's initial scale. However, their data did not help resolve the question of whether polarity reversals occurred at regular intervals or at random.

Both teams were troubled by a report by Sherman Grommé and Richard L. Hay of the University of California at Berkeley. Grommé and Hay, in an article, "Magnetizations of Basalt of Bed I, Olduvai Gorge, Tanganyika," in the November 9, 1963, issue of *Nature,* found that a lava bed in Olduvai Gorge had normal polarity at a time when prior timescales indicated that reverse polarity prevailed.

Both the Menlo Park and Australian groups were cautious in their reactions to the Olduvai findings. Initially the consensus was that the previously reported boundaries between normal and reversed periods might need to be revised. To some extent that was true. But the Olduvai finding might have an even greater significance.

Within the year, Cox, Doell, and Dalrymple reviewed all the evidence—correcting some erroneously dated specimens that they had discussed in previous reports—and concluded that polarity reversals occurred at irregular intervals. They announced their conclusions in the paper "Reversals of the Earth's Magnetic Field," which appeared in the June 26, 1964, issue of *Science.* While subsequent research contributed to the revision of the magnetic reversal timescale, the utility of the scale as a dating tool was no longer in doubt.

The magnetic timescale published by the Menlo Park group in June 1964 was revolutionary for other reasons. Earlier timescales referred to periods of normal polarity with the letter *N* and periods of reversed polarity with the letter *R*. Numbers were used to indicate the age of a period of magnetization. For example, N1 would be the most recent normal episode; N3 would be an older normal period. It got more and more difficult to keep track of all the Ns and Rs and numbers as the timescale was revised. Some means was needed to simplify the terminology.

Cox, Doell, and Dalrymple proposed to call long intervals of a given magnetic polarity "epochs." Short-term reversals would be named "events." In their new timescale, a period of normal polarity that extended from the present back to 1 million years ago was called the Brunhes epoch. The period of reversed polarity from 1 million to 2.5 million years ago was named the Matuyama epoch, and a period of normal polarity from 2.5 million to about 3.4 million years ago was labeled the Gauss epoch. Two events, the normally polarized Olduvai event—discovered by Grommé and Hay—at about 1.9 million years ago and the reversely polarized Mammoth event at about 3 million years ago, were also included.

PRIOR TO THE United States's entry into World War II, Victor Vacquier, a researcher at Gulf Research and Development Company in Pittsburgh, began designing an instrument that, while being towed from an airplane, could detect enemy submarines lurking beneath the surface of the sea. His device, called a magnetic airborne detector, was quickly put into service by Allied forces. The magnetic airborne detector, when installed on a well-armed, long-range B-24 Liberator, proved lethal to German U-boats during the Battle of the Atlantic. (The magnetic airborne detector's first sea trial was on October 21, 1941, when a navy PBY Catalina, using the instrument, found the *S-48*—the submarine that Felix Vening Meinesz and Harry Hess had used for a gravity expedition in 1932—in the waters off the southern coast of New England.) Shortly after the war, a fluxgate magnetometer based on Vacquier's design was developed that could be towed behind a ship. In 1948 Maurice Ewing began taking one of these primitive—and balky—instruments on all his cruises, obtaining measurements of the strength of the magnetic field whenever the machine deigned to work.

During a coffee break at the 1952 meeting of the University of California Institute of Geophysics in La Jolla, California, Ronald Mason asked if anyone had thought of towing a magnetometer behind a ship as had been done with airplanes (he was unaware of the Lamont work at the time). Roger Revelle, then director of Scripps, overheard Mason and asked if Mason wanted to give it a try. Mason agreed.

After searching for a suitable instrument, Mason learned of Lamont's magnetic research at sea, borrowed the Lamont magnetometer, and used it on Scripps's Capricorn Expedition to the South Pacific in 1952 and 1953. Even though the magnetometer was somewhat balky, the results were promising enough to inspire Mason, with the assistance of Arthur Raff, to develop a better instrument.

Early in 1955 Mason learned that the United States Coast and Geodetic Survey was planning a detailed survey of the northeastern Pacific off the west coast of North America. He and Raff contacted the survey to obtain permission to tow their magnetometer from the survey's ship, *Pioneer.* In time the required permissions were granted, and Mason and Raff participated in twelve cruises during 1955 and 1956. After a lot of study of data they had trouble deciphering, Mason and Raff eventually discerned a number of long, linear anomalies of strong and weak magnetism on the ocean floor. When the earth's general magnetic field was subtracted from the measurements, the regions of strong magnetism were found to be positively magnetized, while the areas of weak patterns were found to be negatively magnetized. When mapped, the magnetic anomaly patterns looked like a series of stripes on the seafloor. They were offset along two known east-west faults—the Mendocino and Murray faults—and along one that had been discovered during the 1955 and 1956 surveys and thus named for the survey ship, the Pioneer fault. Mason determined that the offset along the Murray fault was about 84 nautical miles. Vacquier, who joined Scripps in 1956 after Mason was tapped to head an International Geophysical Year research project in the equatorial Pacific, continued the magnetic surveys to the west along both the Mendocino and Pioneer faults and found the offset along the Pioneer fault to be 140 nautical miles, while the Mendocino fault was offset by 640 nautical miles. The magnetic anomaly patterns were found as far to the west in the Pacific basin as magnetometers had been towed, but they disappeared at the continental shelves.

The maps of the anomaly patterns were impressive, but most people—Mason, Raff, and Vacquier included—had no idea what to make of them. Vacquier, Raff, and a third coauthor, Robert E. Warren, suggested in 1961 that the offsets might be evidence of some kind of continental displacement. Specific stripes in the pattern looked as if they had broken apart and moved—as if the seafloor itself had fractured and moved. Noting previous work by supporters of continental drift and of the expansion of the earth—work showing how the continents could have once fit together when you lined up specific features—Vacquier and his collaborators prophesied, "Since it is most unlikely that the magnetic or the physiographic correlations are accidental, the resolution of the paradox will be an important advance in geophysics."

Unfortunately, the prophets failed to take notice when the advance arrived.

FRED VINE became interested in science, particularly geophysics, as a high school student in West London in the years following World War II. It was an exciting time: The International Geophysical Year (1957–1958) was underway in a massive, global effort to learn more about the earth and how it functioned. Vine, who excelled at mathematics, physics, and geography, had the skills to pursue his interest when he entered Cambridge University near the end of the decade.

While working on a natural sciences degree at Cambridge, he began taking a battery of geology courses, including classes by Bullard and Maurice Neville Hill. Hill was a marine geophysicist who, a year after Marie Tharp quietly began mapping the Mid-Atlantic Rift Valley, began surveying the rift while on an expedition of the second HMS *Challenger* and the HMS *Discovery*—the latter originally built for Robert Falcon Scott's first expedition to Antarctica. Thus, Vine was well informed about the latest developments in marine geology and geophysics in January 1962 when Harry Hess came to town to speak on seafloor spreading at a conference on the evolution of the North Atlantic. A couple of months later, in March, Vine heard a presentation by Blackett on paleomagnetism and continental drift.

Since January, Vine had been thinking extensively about Hess's ideas on seafloor spreading. During his final year as an undergraduate he served as the president of the Sedgwick Club, a geological club at Cambridge, and had to give a presentation in May. He chose seafloor spreading as the subject of his talk, "HypotHESSes," because of Hess's recent visit, because of his own interest in the hypothesis, and because research on the topic would help him prepare for upcoming exams. During the discussion someone asked Vine what seafloor spreading had to do with the large areas of linear magnetic anomalies that had been mapped in the northeastern Pacific Ocean by Mason and Raff. Vine's gut feeling was that the two phenomena—seafloor spreading and magnetic "stripes"—were related, but he wasn't sure how.

The summer after he graduated, Vine joined an Arctic expedition to Spitsbergen. Upon his return to Cambridge in October 1962 to begin graduate studies in the Department of Geology and Geophysics, he was assigned as a research student to work under Drummond Hoyle "Drum" Matthews.

Matthews wasn't much older than Vine. In order to fulfill a national service requirement, he had served as an officer in the Royal Navy. Afterward he obtained a natural sciences degree from Kings College, Cambridge, and from 1955 to 1957 he worked as a geologist with the Falkland Islands Dependencies Survey, now the British Antarctic Survey. Matthews—who participated in three expeditions to Coronation Island—later blamed the high-fat diet needed by polar explorers for ruining his health. He returned to Cambridge in 1958 as a doctoral student of Hill's, studying basaltic rocks dredged from the Mid-Atlantic Ridge. Matthews—who had probably been at sea in the Indian Ocean during Hess's visit—had first learned of the seafloor spreading hypothesis when he attended Vine's "HypotHESSes" talk the previous May.

When Vine began his graduate studies, Matthews was again at sea. Hill, who advised Vine during Matthews's absence, decided that Vine should do something with the magnetic anomalies from the ocean floor. Vine studied the literature to find a way to analyze the overwhelming volume of magnetic data that had been collected. He decided to model the anomalies in two and three dimensions using computers. Upon Matthews's return from the Indian Ocean in January 1963, Vine had some new data to work with.

Matthews had conducted a magnetic survey of the Carlsberg Ridge with a magnetometer that he had to deploy by hand. Vine soon began to model the magnetic profile of the ridge and—like Mason, Raff, and Vacquier in the northeastern Pacific—found long linear stripes, some positively polarized, some negatively so, that ran parallel to the long axis of the ridge. Within weeks Vine came up with a hypothesis of his own. He and Matthews discussed it and began writing a paper on it in May. They submitted the paper, "Magnetic Anomalies over Oceanic Ridges," to *Nature.* Vine's idea, bolstered by Matthews's data, breezed through peer review and appeared in September.

The main idea of the Vine and Matthews paper is found on the second of its three pages. "The theory is consistent with, in fact virtually a corollary of, current ideas on ocean floor spreading and periodic reversals in the earth's magnetic field," they wrote. "If the main crustal layer . . . of the oceanic crust is formed over a convective up-current in the mantle at the centre of an oceanic ridge, it will be magnetized in the current direction of the earth's field. Assuming impermanence of the ocean floor, the whole of the ocean crust is comparatively young, probably not older than 150 million years, and the thermo-remanent component of its magnetization is therefore either essentially normal, or reversed with respect to the present field of the earth. Thus, if spreading of the ocean floor occurs, blocks of alternatively normal and reversely magnetized material would drift away from the centre of the ridge and parallel to the crest of it."

A CANADIAN GEOPHYSICIST, Lawrence Whitaker Morley, independently developed a similar hypothesis at the same time. Morley had spent years studying the magnetic properties of rocks and was a pioneer in airborne magnetic surveys. By 1963 he was chief of the Geophysics Division of the Geological Survey of Canada and had little time to conduct his own research. Nevertheless, he submitted his ideas in a letter to *Nature* in January 1963. Unfortunately, probably because he lacked any data to back them up, his very short note was rejected two months later—the journal claimed a lack of space!

Morley resubmitted it to the *Journal of Geophysical Research* in April but heard nothing from the journal until September, after the Vine and Matthews paper had appeared in, of course, *Nature*! At least the *JGR* editor managed to send the note to an anonymous reviewer, but, as Morley recalled, the response was, "such speculation makes interesting talk at cocktail parties, but it is not the sort of thing that ought to be published under serious scientific aegis."

Morley's note begins with a discussion of the linear magnetic anomalies discovered by Mason and Raff in the northeast Pacific. It then goes on to say:

> If one accepts in principle the concept of mantle convection currents rising under ocean ridges, traveling horizontally under the ocean floor and sinking at ocean troughs, one cannot escape the argument that the upwelling rock under the ocean ridges, as it rises above the Curie point geotherm, must become magnetized in the direction of the earth's field prevailing at the time. If this portion of the rock moves upward and then horizontally to make room for new upwelling material, and if, in the meantime, the earth's field has reversed, and the same process continues, it stands to reason that a linear magnetic anomaly pattern of the type observed would result.

In time Morley would receive some long-overdue recognition. He finally got the chance to express his ideas in a presentation in June 1964 at a meeting of the Royal Society of Canada in Quebec City. An article, "Paleomagnetism as a Means of Dating Geological Events," coauthored by André Larochelle, was published later that year as a chapter in the book *Geochronology in Canada*. Finally, the text of his original letter to *Nature* was reprinted in an article, "Canada's Unappreciated Role as Scientific Innovator," written by John Lear for the September 2, 1967, issue of the magazine *Saturday Review.* The hypothesis independently developed by Morley and by Vine and Matthews subsequently became known as the Vine-Matthews-Morley hypothesis.

THE REACTION TO Vine and Matthews's paper was more negative than anything. Vine was at first shocked that most of the magnetic experts, such as at Scripps or at Lamont, did not rush to support the hypothesis. Vacquier, seemingly oblivious to his earlier prophecy, as late as 1964 did not believe that the ideas presented in the Vine and Matthews paper could account for all of the linear magnetic anomaly patterns observed in the world's oceans.

Later, however, Vine understood why the reaction to his hypothesis was less than enthusiastic.

"It was the classic lead balloon, . . . one can't overemphasize that," Vine said. "It was understandable in many ways because it wasn't convincing, particularly if you weren't that close to the data. . . . We didn't know the reversal timescale, so you couldn't correlate it with the reversal timescale. . . . We couldn't document the symmetry about ridge crests, because the data that we had from the North Atlantic and the Indian Ocean just weren't good enough. The spreading rate is low, and we now know that with a low spreading rate the process of formation of crust is more diffuse. The anomaly pattern is not as clearly written as it is in the Pacific, where the spreading rates are high. All that came later. And, of course, basically it put together three ideas which were unproven: One was seafloor spreading, one was reversal of the earth's magnetic field, and one was the importance of remanent magnetization on the ocean floor—that the ocean floor was a really good carrier, that it did make a really good tape recorder."

DOC EWING was as puzzled by the significance of the magnetic data as were Mason, Raff, and Vacquier at Scripps. In 1959 the physicist James Ransom Heirtzler, who had been working with magnetic data as part of an antisubmarine research effort at General Dynamics Corporation, needed to find employment in the New York City area, and Lamont seemed inviting. On a Sunday morning he met with Ewing, who as usual was hard at work at Lamont, to discuss job possibilities. (The three previous times Heirtzler had met Doc were on Sunday mornings while Ewing was at work—one of the three also being Christmas Day!)

Ewing offered Heirtzler a job running the magnetics program. For years Lamont researchers had been towing fluxgate magnetometers behind Lamont's ships. Fluxgate magnetometers such as the ones towed by Lamont and Scripps were difficult to use and had a tendency to break down. Lamont staff had been working on an improvement when Heirtzler arrived. He immediately put his training in nuclear physics to good use and oversaw the development of an improved instrument, the proton precession magnetometer.

The proton precession magnetometer measured the absolute strength of the magnetic field, making unnecessary the laborious calibration calculations that had hobbled Mason, Raff, and Vacquier in their attempts to understand the magnetic patterns observed on the Pacific Ocean floor. The new magnetometer also recorded its data digitally, which made it easier, and faster, to store measurements and to retrieve and analyze the results with the aid of computers, which were becoming more and more common. By 1960 Lamont, a pioneer in using computers for data storage and analysis, was able to obtain a vast amount of magnetic data quickly and in different parts of the world.

New equipment and techniques were not all that Heirtzler helped bring to the magnetics program at Lamont. Late in 1963 Heirtzler hired Neil D. Opdyke, a New Jersey native who had spent much of his life abroad—either in Great Britain or in the British Commonwealth—to set up a paleomagnetics lab. Opdyke, who moved to Lamont from Africa in January 1964, was primarily brought in to fulfill Ewing's wish to have someone responsible for studying paleomagnetism in the vast collection of deepsea sediment cores Doc was so obsessed with. Opdyke was the odd man in the Lamont assemblage in that he, a former student of Runcorn's, was a firm believer both in continental drift and in magnetic reversals.

On *Vema*'s first around-the-world cruise in 1959 and 1960—in which the ship became the first to circumnavigate the globe as well as cross the Arctic and Antarctic circles in the same year—Lamont researchers discovered distinct magnetic anomalies around the Reykjanes Ridge southwest of Iceland. The patterns were similar to those discovered in the Pacific. After the *Vema* cruise, Heirtzler got Victor Goldsmith to continue the magnetic survey of the ridge by air using a Naval Oceanographic Office plane. While that survey was proceeding, Fred Vine and Drum Matthews's paper on magnetic anomalies appeared in *Nature.*

As Heirtzler and his research team began analyzing the Reykjanes Ridge data, they found that the magnetic anomaly patterns were linear and symmetric about the ridge axis—characteristics that followed from Vine-Matthews-Morley hypothesis. Heirtzler remained skeptical. As late as November 1965 Heirtzler, Xavier Le Pichon, and J. Gregory Baron were writing that the ideas presented in Vine and Matthews's 1963 paper did not adequately explain all the magnetic anomaly patterns observed along the ridge. (Morley's paper was not cited in the Reykjanes Ridge article.) Instead, Heirtzler and his coauthors proposed that the anomalies were most likely created by intrusions of lava into cracks on either side of the ridge axis.

Between 1963 and November 1965, when the Reykjanes Ridge paper was submitted to the journal *Deep-Sea Research,* Heirtzler had repeatedly criticized the Vine-Matthews-Morley hypothesis. He was inspired, however, to begin a systematic effort to correlate magnetic anomalies on the seafloor with the magnetic reversal timescale. Believers or not, the Lamont group was rapidly acquiring more data that could be used to put Vine-Matthews-Morley—and ultimately seafloor spreading and continental drift—to the test.

Revelation

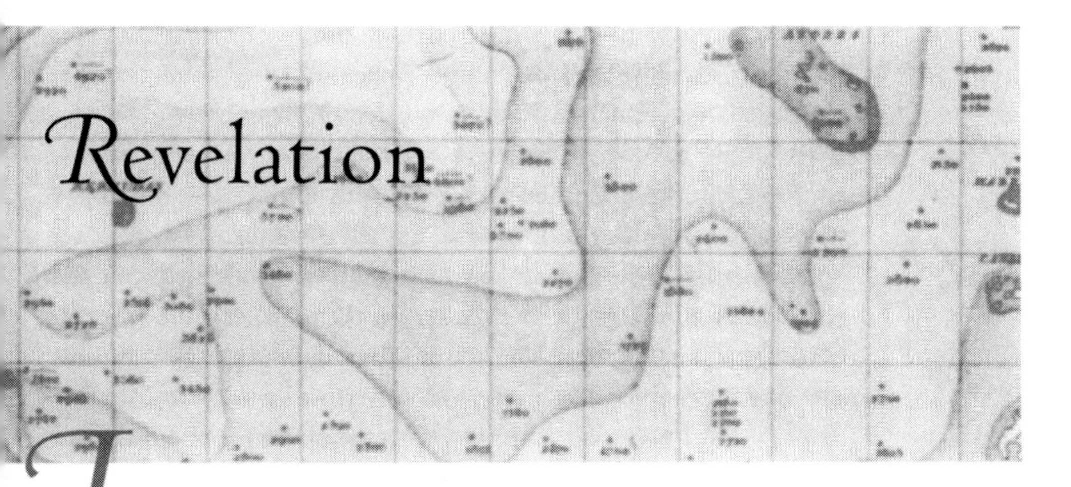

John Tuzo Wilson was a big man—in many ways—with what Fred Vine described as "an incredible physical and mental stamina." He was tall, robust, with a large head, bushy eyebrows, a friendly manner, and a booming storyteller's voice that would never lull a listener to sleep. Wilson was an adventurer, a Canadian Davy Crockett who once killed a moose with an ax—although he denied doing so from the back of another moose, as legend has it. He was a showman who delighted in shocking people, as he would when he sailed his Hong Kong–built Chinese junk, *The Mandarin Duck,* through the frigid waters of Lake Huron's Georgian Bay.

Wilson once summed up his career by saying, "I enjoy, and have always enjoyed, disturbing scientists." Beginning in 1963, shortly after the publication of Harry Hess's "History of Ocean Basins" and Fred Vine and Drummond Matthews's "Magnetic Anomalies over Oceanic Ridges," he would have ample opportunity to do so.

WILSON'S FAMILY BACKGROUND encouraged a sense of adventure and inquiry. His maternal grandfather, Henry Atkinson Tuzo, was of Huguenot descent, born in Trinidad, raised in Virginia, and eventually settling in Montreal. Henry Tuzo, after studying medicine at McGill University in Montreal, joined the Hudson's Bay

Company and served on an expedition to the Canadian Rockies by Sir George Simpson, who, as a high official in the company, was in charge of a vast wilderness territory.

Wilson's mother, Henrietta Laetitia Tuzo, was herself a pioneering mountain climber, making several first ascents by a woman in the Rocky Mountains and Selkirk Ranges. On July 21, 1906, she made the first ascent by anyone on Peak Seven, at the time the last unclimbed member of the Ten Peaks of the spectacular Canadian Rockies. The mountain was later named Mount Tuzo in her honor. The mountain, which looms above Moraine Lake in Banff National Park, is west of the continental divide in what is now Yoho National Park.

Later that summer, in Banff, Tuzo met John Armistead Wilson, a Scot working in a nearby cement plant. They married in 1908 in England, then moved to Ottawa, Canada, where he later became deputy minister of naval service and, after that, director of air services for the Department of Transport, where he organized the development of civil aviation in Canada.

John Tuzo Wilson was born the year after they married.

In addition to his mother, Wilson was exposed early on to a number of other explorers. His father was responsible for equipping the Canadian Arctic Expedition of 1913–1918. When Wilson was fifteen his parents sent him to the boreal forest as an assistant to a forestry party. He subsequently worked for three summers as a field assistant to Noel Odell, who was the last to see famed mountain climber George Leigh Mallory and Andrew Irving nearing the summit of Mount Everest before the two disappeared in the Himalayan mists on June 8, 1924.

Wilson entered the University of Toronto in the late 1920s as a physics student. He soon expressed what to some was a shocking interest in geology and said he wanted to change majors.

"I decided that I should prefer a life in the woods unravelling the mysteries of geology to one in the lab where the practice of physics, then in its heyday, seemed repetitious and stuffy, much as I admired the elegance of its theories," Wilson later wrote. "In the autumn of 1927 when I proposed this transfer all the university authorities were shocked. Professor, later Sir, John McLennan was dismayed and irritated that any promising student should abandon so prestigious a subject as physics for geology, then held in very low regard. Nor did the geologists, who at that time only mapped little patches and identified the rocks and fossils which they encountered, want anyone who might expose their limitations. To their general hostility there was fortunately one exception."

Lauchlan Gilchrist, a geology professor at the university, was involved with a study of the feasibility of using seismic methods in mineral exploration. He was on the lookout for students when Wilson requested a transfer and arranged for the young man to major in both geology and physics. Wilson in 1930 earned the first bachelor's degree in Canada in geophysics.

Wilson next won a scholarship to Cambridge, where he studied under Sir Harold Jeffreys. "I dutifully attended Jeffreys' eight lectures. Unfortunately I could neither hear nor understand them, but Jeffreys was accustomed to this and did not hold it against me," Wilson later said.

He learned more from his tutor, Sir James Wordie, who had been a scientist on Sir Ernest Shackleton's ill-fated but heroic *Endurance* expedition to the Antarctic. "I saw him often, not always for the best of reasons," Wilson wrote, "but he patiently interested me in the joys of exploration, and did not seem upset that I spent my two years in Cambridge in travelling, rowing, flying and drinking."

Returning to Canada at the height of the Depression, Wilson could find no work and was advised to obtain a Ph.D. in geology. He applied to Harvard, the Massachusetts Institute of Technology, and Princeton, and chose Princeton because it "was the only one to offer me any money, was the only one that proposed to teach any geophysics and was the only one where I knew anybody."

Princeton professor Richard Field, who had been such an important influence on Teddy Bullard, Harry Hess, and Maurice

Ewing, took Wilson under his wing. During his tenure at the university Wilson met both Ewing and Hess. While doing his dissertation research in the Beartooth Mountains of Montana he became the first to climb Mount Hague, a flat-topped mountain 12,328 feet high.

In 1936, with a doctorate in geology from Princeton, Wilson began a job with the Geological Survey of Canada. His first task was to map a large region of Nova Scotia, which he did with the aid of aerial photographs—a pioneering move, but one for which he was criticized by the chief geologist for "cheating!" Wilson later used aerial photographs to prepare maps of the glacial geology and tectonic features of the entire country.

Wilson served with the Canadian Military Engineers during World War II. From 1939 through 1943 he was stationed in England and Sicily. From 1944 to 1946 he was director of operational research at the National Defense Headquarters in Ottawa. While there he organized and led "Operation Musk-Ox," a 3,400-mile trek, using tracked vehicles, from the railhead at Churchill, Manitoba, on Hudson Bay north to the Arctic islands, where one of the party relocated the north magnetic pole, and then back south through Coppermine and the Mackenzie River valley to the railhead in Alberta. Shortly thereafter he participated in the United States Air Force's first flight to the North Pole in 1946. Wilson was a member of several other aerial expeditions to the Arctic. By the time he left the army he had risen to the rank of colonel and had been awarded the Order of the British Empire.

In 1946 Wilson returned to civilian life, succeeding his former mentor Gilchrist as professor of geophysics at the University of Toronto. He began traveling widely—over the years speaking at about two hundred universities in more than one hundred countries—seeing for himself much of the world's geology and storing in his mind an encyclopedia, which he would put to effective use by the beginning of the 1960s.

LIKE MOST North Americans, Wilson was not a believer in continental drift. As he tried to make sense of the geology of North America, where the oldest rocks are found on the Canadian Shield while the youngest are typically found along the margins of the continent, he embraced contraction theory as espoused by Jeffreys. As late as 1959 he was firmly in the contractionist camp. After seeing a paper by Robert Dicke on expansion theory, however, he made a brief stop in the expansionist camp, publishing a paper on the consequences of expansion on the earth, but his intellectual travels were not over.

He began evaluating evidence from the ocean floors that had not been widely available until the late 1950s. He was not all that impressed with convection currents, but they did a better job than anything else of explaining heat flow data. He then considered the fracture zones that had recently been discovered in both the Atlantic and Pacific Oceans, comparing them to features like the Great Glen fault in Scotland as well as to similar faults along the eastern shores of North America. By 1960 he had become a drifter.

With the fervor of a zealot, he set out to convert everyone else. Since he had previously argued from the comfort of the contractionist and expansionist camps, he well knew the weaknesses of both. Armed with boundless energy and countless relationships in the international scientific community, he became a formidable proponent of drift. As Wilson modified drift theory in response to the flood of data coming from the ocean basins, however, he sired the ideas that fulfilled Wegener's decades-old prophecy.

Wilson's first foray into the drift world was a brief comment in *Nature* in 1961 on the seafloor spreading ideas of Hess and Robert Dietz in which he solved a problem with Antarctica, which is surrounded by midocean ridges. The problem is, with all the crust being created on the inside of the ring, what happens to Antarctica? Tuzo proposed that, in the case of Antarctica, all the ridges are in effect migrating to the north, pushing the remaining continents in that direction. His note had an additional advantage in that it explained the northerly drift of the continents found in paleomagnetic studies. Wilson had not entirely abandoned expansion, however, suggesting that a slight expansion of the earth might be needed to keep the continents above sea level.

The following year, in the paper "Cabot Fault, an Appalachian Equivalent of the San Andreas and Great Glen Faults and Some Implications for Continental Displacement," which was published in *Nature,* Wilson noted that drifters hurt their cause "by their too rigid concepts of the proto-continents of Pangaea or Laurasia. If, instead of believing that the present continents had all been confined in neat packets before Carboniferous time, they had assumed that continents and fragments of continents had *always* had a random distribution and motion, now joining together in one place and now rifting and separating elsewhere, they could have explained the good fits of some coasts without the necessity of introducing hypothetical and unnecessary distortions" (emphasis added).

Wilson saw that the Cabot Fault in Canada was of similar age as the Great Glen Fault in Scotland. In one of many flashes of insight, he deduced that the two were once joined, when a proto-Atlantic Ocean disappeared in a pre-Pangaea collision between the landmasses that are now Europe and North America. When the two continents began drifting apart after lingering for a while in Pangaea, parts of what had been in North America stuck to Europe; likewise some of what had been in Europe stuck to North America. Thus, rather than embracing the idea that the tectonic history of the earth began with one giant supercontinent that subsequently broke into smaller continents in a one-way process, as many proponents and opponents of drift assumed, Wilson was saying that the process of drift had been going on essentially throughout earth's history, with the continents slamming into one another and tearing apart and with associated closings and openings of ocean basins, in an endless cycle. (He elaborated on this in another paper, titled "Did the Atlantic Close and then

Reopen?" in *Nature* in 1966.) The periods of opening and closing were later referred to as Wilson cycles.

Wilson then added another mechanism to explain the unusual distribution of some islands. He had obtained a research contract from the United States Air Force to compile geological data on the islands of the world. As a result of the work, he focused on oceanic islands in a trio of papers published in 1963: "Evidence from Islands on the Spreading of Ocean Floors" in *Nature*, "Pattern of Uplifted Islands in the Main Ocean Basins" in *Science*, and "A Possible Origin of the Hawaiian Islands" in the *Canadian Journal of Physics*. One thing he noticed was that, in most cases, islands get older with increasing distance from mid-ocean ridges, which is consistent with the idea that they form at the ridges and drift away from them as new crust is created. The pattern he observed was strongest among islands of volcanic origin in the Atlantic and Indian Oceans. The Pacific Ocean proved a problem, however. Many islands followed the pattern he observed, but some, like the Hawaiian Islands, seemed to have no relationship with midocean ridges. In analyzing the data, however, Wilson managed to come up with an explanation of their genesis that was consistent with the theory of continental drift.

Despite his training as a physicist, Wilson did not think in terms of mathematical formulas—in that, he was similar to Alfred Wegener—but instead thought in terms of pictures. A somewhat strange vision led him to his hypothesis on Hawaii's origin. Wilson imagined himself lying on his back in a stream, looking up toward the surface and breathing through a straw bent downstream by the force of the current. This led to his proposal that, as the Pacific seafloor crust drifts across a hot spot in the mantle, magma erupts through to the surface, creating a linear chain of oceanic islands. The youngest, and most volcanically active, island would be directly over the hot spot. The age of the other islands—and seamounts—would increase with distance from the magma source. The youngest island in the Hawaiian chain is Hawaii itself, at the eastern end of the archipelago (although a seamount is already forming to the east of the big island). The island chain extends west-northwest through Midway Island to Kure Atoll and then, in a sudden lurch to the north, continues as the Emperor Seamounts.

The remarkable insights Wilson added to the debate on drift during the first three years of the 1960s would have looked impressive on any geologist's list of lifetime publications. But Wilson was just warming up.

BOTH WILSON AND HESS spent part of 1965 on sabbatical leave at Cambridge, where they had the opportunity to work with each other as well as with Fred Vine and Drum Matthews. Wilson and Hess had been assigned to the same office. Wilson arrived first and quickly expanded his operation to fill the available space. When Hess arrived, he stood in the door, quietly dragging on his cigarette and surveying the extent of Wilson's sprawl, then turned and found somewhere else to work.

Wilson, while revising a book, writing two other papers by himself and one more with Vine, was pondering the major structures of the earth's crust and the forces that act to create and destroy it. Wilson considered the usual large features—ridges and trenches—and one more, the as yet unexplained fracture zones on the deep seafloor. He discussed some of what he was thinking with Hess, Matthews, and Vine, but no one was prepared for the bomb he was to drop—again in *Nature*—in 1965.

As in the case of hot spots, Wilson was inspired by simple images. This time he was playing with paper models in his office in Cambridge when he came to realize that the earth's crust is divided into large, rigid segments called plates. He then predicted three types of borders between plates: midocean ridges between plates spreading apart, mountains or trenches (which in his view were equivalent) between plates colliding head-on, and large, horizontally moving faults between plates sliding past each other sideways—what Scripps and Lamont researchers had been calling fracture zones. Wilson called them transform faults because, at their ends, they are transformed into either a ridge or a trench. These faults, in his view, abruptly end at an intersection with either a ridge or a trench. Three types of transform faults are thus possible: ridge-ridge transforms, trench-trench transforms, or ridge-trench transforms. Wilson predicted that earthquakes along transforms would be confined to the segment between the ridges and/or rifts at either end and that the plates on either side of the faults would be moving in opposite directions.

Wilson's paper, "A New Class of Faults and Their Bearing on Continental Drift," was swiftly published in *Nature*—within weeks of his submitting it. In fact, the paper was moved through the system so quickly that the editor informed him of its acceptance with a phone call rather than the usual letter.

(Incidentally, Alan Melvill Coode, who was working with Keith Runcorn in the Department of Physics at the University of Newcastle, submitted a short note to the *Canadian Journal of Earth Sciences* that also described ridge-ridge transform faults—although he did not refer to them as such. The paper was received on March 30 but did not appear until August of that year—weeks after Wilson's contribution. Even if Coode's paper had been published first, it is unlikely that it would have had the impact of Wilson's.)

The immediate reaction to Wilson's article was unusual for any paper bearing upon continental drift—it was positive.

Wilson was on a roll. During the rest of 1965 he published two additional papers relating to transform faults as well as another with Vine on magnetic anomalies ("stripes" of normal and reverse polarization) and transform faults along the Juan de Fuca Ridge, in the Pacific Ocean west of Vancouver Island. The data in the latter paper were much more convincing than those previously presented in Vine and Matthews's original paper. Wilson and Vine's data clearly showed the symmetry of the magnetic anomaly patterns about the ridge axis, suggesting that crust was continually forming along the ridge and spreading in opposite directions away from it, but their presentation wasn't as convincing as that in

James Heirtzler, Xavier Le Pichon, and Gregory Baron's paper on the Reykjanes Ridge a few months later. The difference was that Vine and Wilson basically showed the anomaly patterns in only one dimension—perpendicular to the ridge axis—while Heirtzler and his coauthors used a two-dimensional format like a traditional map.

Vine and Wilson had the benefit of a more accurate timescale by Allan Cox, Richard Doell, and Brent Dalrymple with which to date the observed anomalies. (Heirtzler and his coauthors also had time to incorporate the new timescale into their analysis but did not, possibly because they still doubted that the earth's magnetic field could reverse itself.) Unfortunately for Vine and Wilson, thin bands of reversely polarized rock on either side of the ridge—which, according to the most recently published Menlo Park timescale, should not have appeared more recently than the Brunhes/Matuyama boundary one million years ago—were too close to the ridge axis and hence too young when compared to the spacing between the more distant, older magnetic reversals. Vine and Wilson were forced to assume that the spreading rate at the Juan de Fuca Ridge had varied over time, slowing in recent times. The assertion was difficult for many geologists to accept. As a result, the paper still was not persuasive enough to convince the skeptics that the Vine-Matthews-Morley hypothesis and seafloor spreading were real.

Wilson's string of hits came to an end toward the end of the year. His final paper of the year described—invented—something called a triple junction, in which three plates come together at a point. *Nature* rejected the manuscript. What a pity—by the end of the decade the existence of triple junctions was well established.

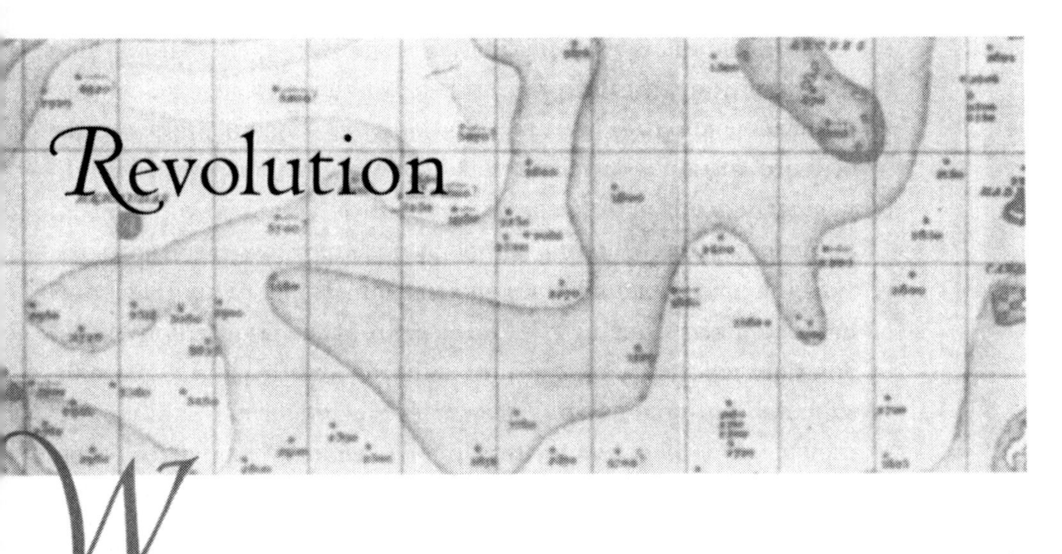

Revolution

Walter Clarkson Pitman III was tiring of his electronics business in 1960 when he decided to go to graduate school to study oceanography. He had earned an engineering degree from Lehigh but had been out of school for several years before he applied to Lamont Geological Observatory. He wasn't accepted right away; instead he was told he would have to go to sea for a year. The problem was not that Lamont officials doubted his qualifications; rather, they needed technicians with solid engineering skills on board the ships. Pitman signed on as a technician aboard the *Vema* and went to sea for nine months.

On his first cruise, Pitman was put in charge of the magnetometer. The assignment was fortuitous. He became fascinated with magnetic studies as well as marine geophysics. Rather than continue with his original plan to become a physical oceanographer—one who studies water chemistry and motion such as tides and currents—when he began his graduate studies at Lamont in 1961, he instead chose to concentrate in geophysics.

"When I went to sea as a technician I was just making magnetic measurements at sea," Pitman said. "I had no—really, very little idea what the object was. I knew that you could use the magnetics data to determine whether bodies were highly magnetized or not."

Pitman was not aware of the debate over the causes of the magnetic anomalies. To him, it seemed that the primary reason for making the magnetic measurements was Doc's obsession with collecting data.

"All that seemed to be of interest to Ewing about the magnetics was that he noticed that over the axis of the midocean ridge he would get a big magnetic anomaly," Pitman said. "He had a philosophy about gathering data, which was gather it while you can, whether you know what to do with it or not, because someday you might find it useful."

Despite his studies, Pitman continued to spend quite a bit of time at sea. When he was back at Lamont, he had a significant amount of work to do that was unrelated to his studies. Because of the hectic schedule both at sea and at Lamont, he found it difficult to keep up with the literature. For instance, he was on board the *Eltanin,* a National Science Foundation–owned ship from which Lamont researchers conducted geophysical surveys of the Southern Ocean, when Fred Vine and Drum Matthews's paper appeared in *Nature* in 1963. Two years would pass before he became familiar with their work, which was relevant to his own research goals: to study magnetic anomalies along fracture zones, in particular the Eltanin Fracture Zone in the Southern Ocean between the southern tip of South America and New Zealand. Pitman was also unaware of some of the adverse comments James Heirtzler and other researchers at Lamont had made about the Vine-Matthews-Morley hypothesis in their own published work.

Among Pitman's many jobs was to evaluate printouts of the data as soon as they were made available. In early December 1965, shortly after *Eltanin*'s twentieth cruise, he was handed a sheaf of paper by Olga Proserowsky with magnetic data from the cruise. Pitman had just read two back-to-back papers in *Science* written by Tuzo Wilson and by Vine and Wilson on magnetic anomalies along the Juan de Fuca Ridge. He had also recently read Robert Dietz's 1961 paper on seafloor spreading and was by now somewhat familiar with Vine and Matthews's 1963 paper. While all the information was fresh in his mind, he realized that he could see in the *Eltanin* data a symmetrical pattern of magnetic anomalies from either side of the Pacific-Antarctic Ridge. Near the ridge axis, the patterns matched what Vine and Wilson had reported from the Juan de Fuca Ridge. Pitman was amazed.

"I went around to several people and showed it to them," Pitman said. "Some just kind of scoffed, but I showed it to Heirtzler, who was my boss, and he was really impressed. And then we together began to show it to quite a few other people, inside the observatory and outside the observatory."

Heirtzler was reluctant to accept that the profile confirmed the Vine-Matthews-Morley and seafloor spreading hypotheses. For a few weeks he argued that some other factor must be responsible for the patterns observed. Neil Opdyke, who for some time had been Lamont's sole advocate of continental drift, and Geoffrey Owen Dickson, a graduate student from Australia who shared an office with Pitman, argued otherwise. Heirtzler's staunch resistance to seafloor spreading and continental drift began to fade as he listened to their arguments and digested the data from the *Eltanin* cruises.

Pitman and fellow graduate student Ellen Mary Herron continued with analysis of the magnetic data from the *Eltanin*-19 and *Eltanin*-21 cruises. The data from *Eltanin*-20 and *Eltanin*-21 were good, but *Eltanin*-19 was different. The bilateral symmetry about the ridge axis was perfect—for about four hundred miles on either side of the ridge. The *Eltanin*-19 data became known as "Pitman's magic profile."

One magic profile might be enough to win a few arguments, but—if Heirtzler's initial resistance was indicative—more would be needed to turn the tide of scientific revolution. Early in 1966 Pitman was preparing to give a talk on the magnetic profiles at the April meeting of the American Geophysical Union in Washington, D.C. He stayed awake one night working up data from additional legs of the *Eltanin* cruises.

"One night I ran out these profiles . . . four or five profiles across the ridges in the Antarctic and Indian Ocean," Pitman said. "And there it was, you could see the correlation between them all. I pinned it up. It was six o'clock in the morning when I got through.

"I put them all up on Opdyke's door and said 'Here it is Neil' with a little note saying that it's six A.M., I'm going home, I need to get some sleep, please don't call me, don't wake me up. And I went home. My younger daughter went off to school and my wife went off to work, and I crawled in bed and went to sleep. About ten seconds past nine the phone rang. It was Opdyke. He said, 'Pitman, get your ass in here right now.' So that's what I did. Boy, he was excited."

In each case the magnetic anomaly patterns were symmetrical about their respective ridge. What was more important, however, was that the anomaly patterns were repeated in each location. Where the profile across one ridge had a thin—or short duration—negative anomaly, for example, the other profiles from different locations would likewise have the thin negative band.

"So that was it, the game was up," Pitman said. "Then it was a matter of going around the other oceans to find the same damn thing."

Heirtzler's group went to work analyzing the huge volume of magnetics data collected by Lamont's ships and others over the years. To cut the workload, the world's ocean basins were divided into sections, with a different team responsible for each. Dickson, Pitman, and Heirtzler were responsible for the South Atlantic; Xavier Le Pichon and Heirtzler took on the Indian Ocean; and Pitman, Herron, and Heirtzler analyzed the South Pacific. The entire group would then sum up the results from all the basins. As they did so, they were able to extend the magnetic reversal timescale well beyond the previously reported 3 million to 4 million years, to more than 75 million years.

The results of the magnetic group's synthesis would not be published for two years. In the meantime Fred Vine would get the opportunity to vindicate himself.

VINE'S WORK with Harry Hess at Cambridge led to his being offered an instructor position at Princeton, which he began in September 1965. Shortly after Vine's arrival in the United States, Tuzo Wilson encouraged him to attend the meeting of the Geological Society of America that November in Kansas City, Missouri. At the meeting Vine and Wilson presented papers based on their research on the Juan de Fuca Ridge. While there, Vine met Brent Dalrymple, who told him about some recent revisions to the polarity reversal timescale. In the months since June 1964 the Menlo Park group had determined that the boundary between the Brunhes and Matuyama epochs was closer to 700,000 years ago rather than 1 million. However, they identified a new normal polarity event, which they called the Jaramillo event, at about 900,000 years ago.

Dalrymple's report was good news to Vine. The revised timescale meant that the seafloor spreading at the Juan de Fuca Ridge had occurred at essentially a constant rate after all, rather than the variable rate that he and Wilson had been forced by the previous timescale to assume. Vine knew the news would spell the end of much of the resistance to the Vine-Matthews-Morley hypothesis. He couldn't capitalize on the news right away, however. He was too busy teaching in November and December, and the following January he was at sea in the Caribbean with Harry Hess and Bruce Heezen.

In February 1966 Vine paid a visit to Lamont, which was only a couple of hours' drive from Princeton, to visit Opdyke. He arrived at an exciting time. Pitman had recently demonstrated the striking similarity in the series of magnetic profiles obtained during the *Eltanin* cruises. Meanwhile Billy Glass, one of Heezen's graduate students, and John Foster, one of Opdyke's graduate students, had discovered magnetic reversals in deep-sea sediment cores, and the reversal patterns in the sediments correlated with the patterns found in the magnetic surveys of ocean basins. The correlation proved beyond any doubt that the anomaly patterns in the seafloor were caused by reversals in the earth's magnetic field. Opdyke himself was excited about a new normal magnetic event that occurred at about 900,000 years ago—until Vine informed him that the event had already been discovered and given the name Jaramillo by the Menlo Park research group.

While Vine was visiting, he met with Heirtzler and Pitman, who showed him the *Eltanin* profiles. The characteristically generous Heirtzler later gave Vine a copy of the *Eltanin* and Reykjanes Ridge data. Vine now had three independent confirmations of the Vine-Matthews-Morley hypothesis: the revised timescale with the Jaramillo event, the *Eltanin* magnetic profiles, and the reversal chronologies from the deep-sea sediment cores. When he returned to Princeton he immediately set to work on a paper, "Spreading of the Ocean Floor: New Evidence," for submission to *Science.*

Pitman and Heirtzler were working on their own paper at the time, too. Later that spring, when Heirtzler was out of town, Pitman learned that Vine was to return to Lamont in April or May to give a colloquium talk. He feared being scooped by the talk, so he sent Vine a draft of his and Heirtzler's paper. They were at an impasse. Pitman suggested publishing together, an offer Vine—who had taken quite a beating in the literature, especially from Lamonters, over the Vine-Matthews-Morley hypothesis—justifiably declined. After Heirtzler returned to Lamont, he and Pitman went down to Princeton to discuss matters with Vine. Some wrangling erupted between the Lamont and Princeton groups as well as with the editors of *Science*—Phil Abelson, the editor of *Science* and a friend of Hess, asked Hess if the Vine paper was worth publishing. Ableson was apparently worried that the magazine was carrying too many earth science papers at the time!

Eventually a solution was reached that left everyone satisfied. The Pitman and Heirtzler report, "Magnetic Anomalies over the Pacific-Antarctic Ridge," was published on December 2. Vine's paper, a major review of the recent findings including those published on December 2, appeared two weeks later.

VINE WASN'T the only person to get excited about the *Eltanin* profiles. Lynn Sykes had been trying to understand the behavior of seismic waves that cross ocean basins. Needing accurate data on epicenter locations, he began revising a computer program developed by Bruce Bolt at Lamont to determine the location of earthquakes. Bolt wasn't terribly interested in the application of the program, however. Sykes took it over and, while working on a paper published in 1963, found that the program reduced the probable error in earthquake locations to about six miles rather than hundreds of miles as before. With this more accurate seismic data, he discovered a fracture zone about 300 miles long in the South Pacific.

He realized that work on locating earthquakes was more important than the surface wave work he had been doing and spent the years from 1962 to 1965 discovering new fracture zones in the oceans of the world. Wilson had, in fact, been inspired in part by Sykes's data when he developed the concept of transform faulting. Ever since Wilson's "New Class of Faults and Their Bearing on Continental Drift" paper came out in 1965, Sykes had been thinking of testing the hypothesis, thinking that it would most likely fail the test. For the time being, however, he was preoccupied with a manuscript on earthquakes in the Fiji-Tonga region of the Pacific Ocean.

In February 1966 Heirtzler and Pitman showed Sykes and Jack Oliver, who now directed the seismology program at Lamont, the magic profile. They had the original printout plus an acetate overlay that could be reversed and laid on top of the original to display the symmetry. Sykes reacted immediately.

"I really thought [Sykes] was going to jump out the window and take the short way back to his office," Pitman said. Sykes knew he could test Wilson's transform fault hypothesis by analyzing seismic records to determine the initial direction of movement during an earthquake. If Wilson was right, Sykes would find evidence only of vertical motions characteristic of normal faults along the rift zones, while along the fracture zones Sykes should find horizontal motions characteristic of what had been traditionally called strike-slip or transcurrent faults. The slipping motions would be in opposite directions on either side of the fracture and would be confined to the segment of the fracture between the features it connects (two ridges, a ridge and a trench, or two trenches).

Sykes dropped everything else he was working on to test Wilson's hypothesis. He could not only call upon his earthquake *location* data but also use data from the World Wide Standardized Seismograph Network (in large part built upon the foundation of a global seismic network set up by Ewing, Oliver, Frank Press, and George Sutton) to determine the *direction of movement* along the faults. Sykes analyzed in detail seventeen earthquakes located on midocean ridges or their terrestrial extensions—such as the East Africa Rift Valley. It took only a few weeks of work to settle the issue.

Twelve of the seventeen earthquakes Sykes analyzed had occurred on either the Mid-Atlantic Ridge or the East Pacific Rise. Of those, ten had occurred along fracture zones that intersected the crest of the ridges. Sykes found that seismic activity along fracture zones occurs primarily in the region between ridges and that the earthquake motions were consistent with Wilson's transform fault hypothesis. Furthermore, he concluded that such motions could occur only if seafloor spreading was occurring. Sykes's analysis, rather than disproving Wilson's ideas, decisively confirmed them.

WORD OF Pitman's magic profile and Sykes's confirmation of Wilson's transform fault hypothesis spread throughout Lamont. Pitman and Heirtzler presented a paper on the *Eltanin* profiles, and Opdyke spoke on the analyses of the deep-sea sediment cores, at the spring meeting of the American Geophysical Union in Washington, D.C. As a result a number of vehement opponents of continental drift were becoming converts.

At the time, Paul Gast, a geochemist at Lamont, and Robert Jastrow, director of the Goddard Institute for Space Studies in

New York City, were organizing a by-invitation-only symposium on continental drift. Presenters would discuss their analyses. At the end of the presentations Teddy Bullard would debate the case for continental drift, while Gordon MacDonald would state the case against. The list of invitees was short and highly selective: Other than researchers in the field, only the most influential members of the North American scientific establishment were allowed. They gathered on the morning of Thursday, November 12, 1966, for two days of discussion.

The presentations were organized into four sessions: bulk chemical and physical properties, paleomagnetic and paleoclimatic studies, submarine geology and crustal history, and intercontinental correlations in space and time. Five people presented the new evidence for continental drift from the ocean basins: Opdyke spoke on paleomagnetism in sediment cores, Vine on magnetic anomalies about midocean ridges, Heirtzler on seafloor spreading, Le Pichon on heat flow, and Sykes on seismicity. Bill Menard, who by then had become a convert himself, discussed questions about the ocean basins that remained to be answered in terms of drift theory.

The debate was spirited. For example, Vine ably parried Frank Press, who questioned the method used to analyze the symmetry of the magnetic anomaly patterns.

"Have you tested the symmetry statistically?" Press asked.

"I never touch statistics," Vine answered. "I just deal with the facts."

From the questions included in the symposium proceedings, *The History of the Earth's Crust*, published in 1968, it was clear that everyone familiar with the oceans had become believers in seafloor spreading and continental drift. Those few scientists who still found it difficult to accept drift were specialists in terrestrial geology and paleontology. By the end, however, those who believed in some form of drift prevailed. Bullard confidently declared the matter settled in favor of continental drift. MacDonald, who had been a vocal critic of drift theory, found himself required elsewhere. No one presented antidrift remarks in his place.

"My work and that of Pitman and Heirtzler and others at Lamont on magnetics and heat flow made a great impression on the invited audience," Sykes later observed. "No one had intended the meeting to be a decisive conference on earth mobility, but it turned out to be the turning point for those concepts in the United States."

OLIVER HAD ALSO headed back to his office inspired by his viewing of the magic profile in February 1966. He and another Lamont seismologist, Bryan Isacks, had begun the "Tonga-Fiji Deep Earthquake Project" in the South Pacific in 1964. They were solely interested in deep earthquakes—earthquakes whose foci lie in the mantle below the earth's crust—without any plans to address the theory of seafloor spreading. So far everything that had been discovered could fit either the expansion or the seafloor-spreading theory. The question of how the earth accommodated the crust created at the rifts still needed to be answered.

Deep earthquakes, some as far as four hundred miles below the earth's surface, had been well documented. A Japanese seismologist, Kiyoo Wadati, first demonstrated their existence. Hugo Benioff, a seismologist at the California Institute of Technology in Pasadena, associated these deep earthquake zones with massive faults at the interface between the continents and oceans. The zones eventually became known as Wadati-Benioff zones.

Fiji and Tonga were selected for study because the region experienced the greatest number of deep earthquakes of any place in the world. Isacks moved to Fiji in 1964 to set up and monitor the seismic network for the project. He was first to see the data, and was puzzled by them. High-frequency shear waves from the frequent deep shocks arriving in Tonga were three orders of magnitude greater than those arriving a similar distance away in Fiji. What was damping the signal?

Several months after Oliver saw the magic profile, he and Isacks were in his office sketching on a blackboard a cross section of the zone where the seismic waves traveled faster, with less attenuation. It looked like a slab—a descending slab. Suddenly they realized what their sketch implied. A gigantic slab of oceanic crust about sixty miles thick was being thrust or dragged down into the mantle along the Tonga-Kermadec arc. Their discovery meant that crust created at the rifts is compensated by crust destroyed in the deep-sea trenches. An expanding earth was no longer needed to accommodate the new crust created at the ridges. Despite Warren Carey's unrelenting attempts to defend the expanding-earth theory, seafloor spreading had won.

LE PICHON was a French scientist who had come to the United States to study at the invitation of Maurice Ewing—Doc had called Le Pichon on a Christmas Day to extend the invitation. Le Pichon played an integral part in most of the ocean-floor research going on at Lamont throughout the early 1960s. He was as vehement as nearly everyone else at Lamont in his opposition to continental drift and seafloor spreading. Late in 1965 he finished writing his dissertation. In it, he concluded that convection currents probably do occur in the mantle, but that the seafloors do not spread. He was on a cruise to the South Atlantic throughout the early part of 1966; then he defended his dissertation in Strasbourg, France, in April.

Upon his return to Lamont in May, he saw the magic profile for the first time. Le Pichon did not like what he saw. After he returned home that day, he asked his wife to get him a drink and then dropped a bombshell.

"The conclusions of my thesis are wrong," he said. "Hess is right."

Le Pichon attended the meeting of the American Geophysical Union in Washington, D.C., in April 1967, in which the vast majority of the American earth science establishment became converted to continental drift. On April 19 Hess presided over a

"Seafloor Spreading" special symposium for which more than seventy abstracts had been submitted. W. Jason Morgan of Princeton was scheduled to give a talk on the formation of ocean trenches by convective processes. Le Pichon and Manik Talwani, also of Lamont, were preparing to pounce on Morgan because they had some serious disagreements over the topic, but Morgan ruined their plans by presenting something completely different.

Morgan announced that he was going to do a geometric analysis of the motion of plates—or blocks as he called them—on a sphere. In it he divided the crust of the earth into a number of blocks and applied a theorem attributed to eighteenth-century Swiss mathematician Leonard Euler to the effect that the movement of a rigid block on the surface of a sphere would be a rotation about some axis. Morgan's talk formed the basis for what later became known as plate kinematics.

(Dan McKenzie of Cambridge and Robert Parker of Scripps had independently developed a plate kinematics approach based on Euler. They, however, used a computer program to simulate plate motions. McKenzie, ironically, had been at the AGU session but had no interest in hearing about convection currents and trenches. He skipped Morgan's talk.)

Le Pichon tested Morgan's method by trying to compare it with the magnetic anomalies of the seafloor. He began working on a complicated battery of computer programs that could simulate plate motions. His analysis bolstered the conclusion that the plates were rigid and that the various rift motions obeyed the principles of spherical geometry. Le Pichon next applied the principles to simulate converging motions along the trenches. The results agreed well with seismic data from the trenches. Le Pichon then attempted to apply plate kinematic principles to the reconstruction of past plate motions—he was the first to do so.

The new theory, which had yet to be named, shortly had a solid mathematical foundation that Wegener's and other theories of continental drift lacked. Within a seven-month period in 1967 and 1968, three papers explaining the geometry of plate movements across the surface of the earth appeared. McKenzie and Parker were published first. Their paper, "The North Pacific: An Example of Plate Movements on a Sphere," appeared in *Nature* in December 1967. Morgan's attempt to publish his paper, titled "Rises, Trenches, Great Faults, and Crustal Blocks," was held up by troubles with a reviewer (Menard) and did not appear until March 1968, in the *Journal of Geophysical Research.* Finally Le Pichon's paper, "Sea-floor Spreading and Continental Drift," also appeared in the *Journal of Geophysical Research,* in June.

The revolution was largely complete. The basic components of a new theory—originating with Alfred Wegener but with important revisions by Hess, Wilson, and Robert Dietz, had been assembled. Tests of the theory had been proposed and conducted by Vine and Drummond Matthews; Heirtzler, Opdyke, Pitman, and the rest of the Lamont magnetics group; and Isacks, Oliver, and Sykes in Lamont's seismology department. A mathematical foundation had been developed and applied by Morgan, McKenzie and Parker, and Le Pichon. One more paper—the one to put the major pieces of the puzzle together—was needed to seal the victory.

It came from Isacks, Oliver, and Sykes, who began working in 1967 on a synthesis of what was then known. No one member of the trio dominated the thinking—the order of authors was determined at random. Their paper, "Seismology and the New Global Tectonics," was published in the *Journal of Geophysical Research* in September 1968 and outlined the new ideas about large-scale motions of segments of the earth's crust and about the processes within the earth that drove those movements. The three authors acknowledged the great ideas that formed the backbone of the new theory as well as the prophets who had originated them: continental drift by Wegener; seafloor spreading by Hess and Dietz (making sure to list Hess before Dietz—in a sense fulfilling another prophecy, that the last [published] shall be first [credited]); and transform faults by Wilson. Wegener's vision of the connections between disparate phenomena such as crustal motions, earthquakes, vulcanism, and mountain building was fully confirmed when Isacks, Oliver, and Sykes proclaimed the good news of a new, unifying paradigm in earth science.

"Surely the most striking and perhaps the most significant effect of the new global tectonics on seismology will be an accentuated interplay between seismology and the many other disciplines of geology," the trio wrote in their final paragraph. "The various disciplines which have tended to go their separate ways will find the attraction of the unifying concepts irresistible, and large numbers of refreshing and revealing interdisciplinary studies may be anticipated. . . . Even if it is destined for discard at some time in the future, the new global tectonics is certain to have a healthy, stimulating and unifying effect on all the earth sciences."

Alpine Vistas to Abyssal Plains

Throughout the 1950s until the mid 1960s, Bruce Heezen had helped blaze the underwater trail to the promised land of plate tectonics. But as the spirit of revolution erupted throughout the earth sciences in 1967 and 1968, Heezen could not participate. Rather, he could only view the advances from afar. Rather than fight to advance science, he was embroiled in a bitter struggle for survival at Lamont.

At Lamont, Ewing had established, and commanded, a disciplined team effort. Under his system everyone pitched in to collect the data; thus anyone should have access to it. The system worked fine so long as there was more than enough work to go around and as long as the Lamont Observatory (and, not incidentally, Ewing as its director) received a share of the spotlight. Individual scientists could set up their own fiefdoms over the data they were most interested in—but as the number of researchers increased, conflicts arose.

Once one of Ewing's favorite students, Heezen eventually wanted to make a name for himself. His ambition collided with Ewing's demand for absolute loyalty as well as with the ambitions of other young scientists who, like him, wanted more control over their own scientific destinies. Rather than discussing his concerns with Doc, Heezen began to ignore Ewing's rules. And the troubles began.

Long-festering resentments—the origins of which now seem as mysterious as Edgar Allan Poe's "thousand injuries of Fortunato"—erupted into all-out war following the Second International Oceanographic Congress in Moscow in 1966.

The year before, Billy Glass and John Foster had discovered magnetic reversals in deep-sea sediment cores. They weren't the first to do so—a few researchers, such as Christopher G. A. Harrison, of Cambridge University and Scripps Institution of Oceanography, and Brian M. Funnell, of Cambridge, had also—but the earlier work had been hampered by problems with the magnetic timescale and insufficiently sensitive instruments. By the time Glass and Foster made their discovery, they could identify tiny reversals in the sediments, such as the Jaramillo event, and correlate those with similar reversals in the patterns of magnetic stripes on the ocean floor. Both Opdyke and Heezen realized the significance of the discovery; their squabbling over who should get credit for what erupted into a controversy that nearly tore Lamont apart.

According to Heezen, he and Opdyke initially agreed that Heezen would be a coathor on the paper that announced the Glass and Foster discovery—as long as Opdyke's name was listed first. In time, Opdyke suggested that James Douglas Hays, a former student of Heezen's who had used fossil radiolaria to date the cores Glass and Foster had worked on, be added to the list. Heezen agreed to that. Later, though, Opdyke asked Heezen to withdraw from the initial paper.

"I told Bruce Heezen that he could not be an author on the first paper," Opdyke said, "which caused a big blowup because I'd been doing paleomagnetism—it's my business—and his very presence on the paper would have been enough to lead people to think he had been the prime mover."

Opdyke was just getting started in his career and realized what such a discovery meant to his status in the scientific community. He feared that if Heezen's name appeared on the list of authors, others would think the work was Heezen's idea, and he was not in the mood to share credit with anyone (not even Ewing, who had wanted him to look at the cores in the first place). Therefore Opdyke sought to ensure that either he or Glass would present the research first.

Opdyke did in fact discuss the findings first, at the spring meeting of the American Geophysical Union in Washington, D.C. Heezen later wrote (in some notes documenting his version of events surrounding the controversy) that he was unaware of Opdyke's plans to discuss the sediment core research at the AGU meeting. While Heezen had not attended the meeting, he said he had heard that Opdyke had also discussed the work of Dragoslav Ninkovich, a postdoctoral researcher affiliated with Heezen, on sediment cores from the North Pacific.

Opdyke had also told Heezen that he must not mention the sediment core research at the upcoming oceanographic congress in Moscow. Heezen disregarded Opdyke's instructions and discussed the work on reversals in the sediment cores anyway. Heezen wrote that Ninkovich had already prepared a paper on the North Pacific research prior to the Moscow meeting and, fearing that Opdyke was going to take credit for his work, asked Heezen to present it. After the talk, Heezen was asked to attend a press conference to discuss the work. His comments appeared in an article in the June 2 issue of *The New York Times* and in a follow-up story in the June 17 issue of *Time* magazine. Both articles gave the impression that Heezen was taking credit for the work—that is what his adversaries at Lamont felt.

Heezen himself felt victimized. He claimed the stories resulted from poor translation at the Moscow press conference and from sloppy reporting. *The New York Times* story did at least mention all the members of the research team. (The less said about the *Time* story the better.) Heezen was in Europe when the reports were published. He saw a copy of *Time* in a newsstand in Innsbruck, Austria, read the story, and was horrified. He quickly sent a cable to the editor of the magazine demanding a correction and clarification. By then it was too late, however. His colleagues were already furious at his perceived transgressions.

Opdyke—unaware of the building tensions between Ewing and Heezen—complained to Doc after seeing the article on the Moscow meeting in *The New York Times.* Ewing responded with a fury that could have tamed a West Texas tornado. He tried to fire Heezen but could not do so because Heezen had tenure. He tried to bar Heezen from the Lamont facilities but again could not do so because Heezen, as a tenured professor at Columbia University, had a right to access facilities owned by Columbia.

Ewing, however, did manage to deny Heezen access to Lamont data and Lamont ships, and to prevent Heezen from applying for research funding through Lamont. (Among the applications that were killed was one seeking funds to allow Heezen to get his own spinner magnetometer similar to the one that Foster had built for Opdyke.) Doc seized control of grants in which Heezen was principal investigator and gave the projects to others—a move of questionable judgment that in more recent times could have imperiled all of Lamont's outside research funding.

Ewing also took out his frustrations on Heezen's associates, even though they were only peripherally involved in the conflict. He fired Marie Tharp, who from then on worked from her home funded by grants Heezen secured outside the Lamont hierarchy. One day Ewing told Glass to pack up his things and leave Lamont in two hours. Glass did as he was told. (Doc invited Glass back a couple of weeks later.)

Ewing's efforts to rid himself of Heezen were stalemated, but the feud left Lamont divided and weary—and may have contributed to Lamont's ridding itself of Ewing in 1972 when Columbia University officials forced him into mandatory retirement from his post as director of Lamont. Doc wanted to fight the university's decision but found he did not have enough support among Lamont's researchers to do so. When offered a position as head of the Division of Earth and Planetary Sciences at the Marine Biomedical Institute of the University of Texas Medical Branch in Galveston, he decided to return to his native Texas.

Joe Worzel was among a handful of Lamonters who went to Texas with Doc. Together, as in the early days of Lamont, they began assembling a marine research lab from scratch. The two—without planning to do so—moved into homes in Galveston that were across the street from each other. Ewing lived just long enough to see the results of the first research expedition of the newly formed lab.

At two in the morning on April 28, 1974, Worzel was awakened by a knock at the door. Harriett, Ewing's third wife, said she thought Doc had had a stroke and asked Worzel to help get him to the hospital. Together they struggled to get Ewing out of the bathroom, where he had collapsed, and into the back seat of Worzel's car. Harriett cradled her husband's head in her lap while Worzel sped the five miles to John Sealy Hospital in Galveston. Ewing never regained consciousness and died on May 4.

Ewing was buried in Sparkill, New York, just a few miles from his beloved Lamont. Sadly, Worzel, who had been working with Doc longer than anyone else and who was probably his most loyal associate, was not asked to give the eulogy at the funeral. Frank Press, who was among the first of Doc's former students to leave Lamont—he took a position at California Institute of Technology in Pasadena in 1955—delivered it instead.

HEEZEN AND THARP were well on their way to mapping the world's ocean floor by the time the feud with Ewing erupted. The United States Navy had prohibited them from producing contour maps such as the crude ones prepared by Matthew Fontaine Maury in the 1850s or Sir John Murray in the early years of the twentieth century. The navy had funded the collection of much of the sounding data and wanted that data classified to prevent its release to the public (and, more important, to lurking Soviet submarines). To get around that bureaucratic obstacle, Heezen and Tharp, when they began mapping the ocean floor in 1952, decided to follow the lead of cartographer Amin Kohl Lobeck, who had pioneered techniques of mapping the physical features of landscapes. Instead of a contour map, they would produce a physiographic map—a drawing, without detailed depth information, of the way the ocean floor would look with all the water removed. The navy had no objection to that.

"It allowed us to capture the sea floor's many textured variations, contrasting the smoothness of the abyssal plains, for example, with the ruggedness of the mountains along the ridges," Tharp said. "In retrospect, our choice of map style turned out to be significant because it allowed a much wider audience to visualize the sea floor."

The mapping work was much more challenging than the troubles Tharp had experienced in preparing the North Atlantic profiles. First, she and others working under her supervision plotted soundings along the tracks taken by research and survey vessels on a base map generally covering an area of one degree latitude by one degree longitude. Next, Tharp would convert the sounding lines into depth profiles, much like the original six prepared for the North Atlantic. Afterward she would make three-dimensional sketches of the ocean-floor topography along the sounding lines.

Tharp and Heezen would then interpret the sketches and estimate the topographic trends in areas between the sounding lines. They would interpret the entire one-degree by one-degree chart at least twice, match it against a chart of the entire region or ocean basin, and reinterpret everything, then plot a final chart. The process would be repeated—they often changed their minds—until they were finally satisfied with their mapping of a given portion of the ocean floor.

Neither Tharp nor Heezen was daunted by the massive volumes of data in their possession. Tharp, instead, was haunted by what was missing. "We'd use everything available, but there would be blank areas," she said. "I'd work up a blank area with what we could get. The biggest challenge was just to keep providing data for the blank areas and to upgrade the areas that we changed our minds on."

Tharp resorted to time-honored traditions to fill in large areas for which there were no data.

"Like the cartographers of old we put a large legend in the space where we had no data," Tharp said. "I also wanted to include mermaids and shipwrecks, but Bruce would have none of it."

Heezen and Tharp's first physiographic map, of the North Atlantic, was published in the Bell Telephone System's *Technical Journal* in 1957 and reprinted by the Geological Society of America two years later. The map was a success, so the two immediately began work on a physiographic map of the South Atlantic. The South Atlantic map, which was published in 1961, caused a stir because of the almost perfect conformity in shape of the Mid-Atlantic Ridge to the South American and African coastlines. In the past, many scientists had found it easy to dismiss the near-perfect fit of the two coastlines as evidence of continental drift, but now the appearance of a third feature—the Mid-Atlantic Ridge—similar in shape to and located between the two coasts, shook their skepticism. It seemed highly unlikely that such similarities could have arisen unless South America and Africa had once been joined.

Heezen had next planned to map the Mediterranean Sea, but a major study of the Indian Ocean—the International Indian Ocean Expedition—was in the works, and a map was needed. Heezen and Tharp thus redirected their efforts. By the time the Indian Ocean work got underway, the feud between Heezen and Ewing had begun to affect the mapping efforts. Heezen began collaborating with scientists all over the world—he needed to obtain ship time from other institutions and to secure additional sources for data because he was denied access to Lamont's ships and data. The Indian Ocean map itself was an international effort, with sources in the Soviet Union, Japan, the United Kingdom, Australia, and South Africa, as well as the United States, contributing data. It was published in 1964.

(The map had a glaring error, however. At the time it was published, Heezen and Tharp were obsessed with fracture zones, and no one had published anything about the existence of triple junctions—Wilson's paper on the topic had not been written, nor rejected, yet—and they completely misinterpreted a section of the Indian Ocean where three ridges come together. They later corrected the mistake.)

Heezen and Tharp's Indian Ocean map drew the attention of the editors at *National Geographic Magazine,* who wanted a map of their own for an article on the international expedition. Rather than the pen-and-ink drawing that Tharp had been producing, they wanted to produce a full-color map, something that looked like a painting.

Prior to contacting Heezen and Tharp, *National Geographic* had received a letter from an Austrian girl who wrote, "I've been looking at your maps and my father can paint better than you can." The magazine sent Albert Bumstead, its chief topographer, to Innsbruck, Austria, to meet artist Heinrich Berann.

"Berann had a beautiful home and a separate studio, and he did serious paintings there," Tharp said. "But he couldn't make any money with his serious paintings. So he took up doing mountain panoramas for the Austrian ski tourist trade."

National Geographic teamed Heezen, Tharp, and Berann to produce a new Indian Ocean map, which was published in October 1967. The map was a success, and the trio had enjoyed working together. When *National Geographic* asked them to make maps of the rest of the world's oceans, they agreed. The final map in the series, of the seas around Antarctica, appeared in 1975.

After mapping the individual ocean basins of the world, Heezen, Tharp, and Berann had just one more thing to do. In 1974 they submitted a proposal to the Office of Naval Research for a map of the world's ocean floor. They faced two main tasks: to update the depth data and to simplify the information in order to reduce it to the smaller scale of a world map. They completed their effort early in 1977.

Shortly after approving the proofs and sending them to the printer, Heezen went to sea again, eager for yet another chance to see the ocean floor he had devoted his life to studying. Tharp, who as long as Ewing was at Lamont was prohibited from going to sea on a Lamont ship, was also at sea, in the eastern Atlantic aboard the British research vessel *Discovery.* The two planned to meet in Reykjavík, Iceland, after their respective cruises were over.

Heezen was part of a United States Navy–sponsored expedition to the Reykjanes Ridge southwest of Iceland. He had long been involved with expeditions aboard submersibles and submarines, and this time was aboard the navy's nuclear-powered research submarine *NR-1,* which was being escorted by its support ship, *Sunbird,* on June 21, 1977. Heezen was preparing for a dive to explore the ridge when he suffered a fatal heart attack.

Heezen's body was flown by helicopter to Iceland, then transported by airplane to Norfolk (Virginia) Naval Station, where an autopsy was performed. Heezen and Tharp had long planned to be buried together, but his mother, Esther H. Dauch of Pittsburgh, arrived at Norfolk before Tharp and claimed his body. Dauch had him interred in the family cemetery in Iowa.

Tharp had to oversee the printing of the *World Ocean Floor* map alone. Because of concerns over publication delays, Bill Ryan, a former student of Heezen's who had remained at Lamont as a researcher, was asked to help Tharp complete the process of printing the map. She seemed reluctant to finish it: Letting the presses roll was like letting go of Heezen. Tharp kept adding data points in places—and stopped the presses right after they had started to add one more sounding—but afterward the presses were allowed to run uninterrupted.

Heezen, Tharp, and Berann had produced something much greater than a map of the world's ocean floor. Like a sacred religious icon, the map inspires a sense of awe at the majesty and mystery of the earth. It fills in the vast blank spaces that once reflected what humans did not know about the planet, yet stirs a desire for even more exploration of the watery realm.

The *World Ocean Floor* map and its predecessors also played a vital role in the argument over continental drift and plate tectonics.

"Scientists and the general public got their first relatively realistic image of a vast part of the planet that they could never see," Tharp said. "The maps received wide coverage and were widely circulated. They brought the theory of continental drift within the realm of rational speculation. You could see the worldwide mid-ocean ridge, and you could see that it coincided with earthquakes. The borders of the plates took shape, leading rapidly to the more comprehensive theory of plate tectonics."

More than twenty years later, better technologies have been developed and more accurate maps have been produced, but none of the newer maps will match the influence and emotional impact of the work of Heezen, Tharp, and Berann. The *World Ocean Floor* map opened the eyes of scientists and the public, who have since viewed the earth in an entirely new way.

Summing Up

John McPhee, in his book *Basin and Range,* wrote that if he had to restrict his writing on plate tectonics to one sentence, it would be, "The summit of Mt. Everest is marine limestone."

His sentence deftly captures the importance of plate tectonics theory today. Plate tectonics explains the origin of the earth's major structures, ranging in size from continents and ocean basins down to mountain ranges, rift valleys, and oceanic islands, and in elevation from the highest peaks of the Himalayas to the deepest trenches of the Pacific. The results of plate tectonic processes abound, from the mighty folds of the ancient and well-worn Appalachians to the majestic escarpments of the East African Rift Valley and the Palisades of the Hudson River, and to the volcanoes that imperil cities in Iceland, Indonesia, Japan and Mexico. Idyllic island resorts in Hawaii and the Caribbean owe their existence to the movements of great slabs of the earth's crust. Other island groups, such as Ascension, St. Peter and St. Paul Rocks, Tristan da Cunha and Gough, now lie among the high peaks of the midocean ridges, but will likely disappear as a result of the crust's unceasing motion. The theory explains all volcanism and most earthquake activity on earth. Its importance reaches far beyond the earth sciences, too. For instance, one cannot understand the diversity of life on this planet without some knowledge of both evolution and plate tectonics. In many parts of the world, billions of dollars are spent to prepare buildings, roads, and humans for the inevitable volcanic or earthquake event.

Alfred Wegener realized the far-reaching significance of continental drift—the precursor to plate tectonics. His critics realized it, too, and some, like Rollin Chamberlin, used the potential implications of the theory as an argument against it.

"But taking the situation as it now is, we must either modify radically most of the present rules of the geological game or else pass the hypothesis by," Chamberlin wrote in the paper he presented on continental drift at the 1926 American Association of Petroleum Geologists symposium in New York. "The best characterization of the hypothesis which I have heard was a remark made at the 1922 meeting of the Geological Society of America at Ann Arbor. It was this: 'If we are to believe Wegener's hypothesis we must forget everything which has been learned in the last 70 years and start all over again.'"

Wegener remains a controversial figure today. To some he is a scientific hero, an innovative thinker who ignored convention and criticism while inspiring overthrow of the old order. To others he is an overrated thinker who was wrong more often than right, a scientist who has benefited from revisionist polishing of his reputation.

Some experts dismiss Wegener's effort by stating that plate tectonics and continental drift are two different theories—with the implication being that he should get no credit for plate tectonics. True, continental drift and plate tectonics are different, but drift is the ancestor of plate tectonics. Without Wegener's demonstration of the transience of continents and ocean basins, and without his example of how to solve the mystery of the motions of the earth's crust by harnessing evidence from a variety of scientific disciplines, we might still be puzzling over things like the magnetic anomaly patterns on the ocean floor.

True, Wegener was wrong in significant areas—but he himself typically identified the points now known to be erroneous, such as his proposed mechanism for drift, as problems. Like any good scientist, he knew that some parts of his theory would not stand the test of time.

For decades, a big obstacle to acceptance of Wegener's theory seems to have been his critics' insistence that he prove the cause of drift before they would accept the existence of drift. This seems ludicrous to me. A police officer does not have to know the cause of an auto accident before filing a report stating that the wreck occurred. The dented fenders and broken glass and plastic on the highway are solid evidence that some kind of impact occurred. Whether driver A or driver B or any particular combination of drivers, road conditions, and mechanical malfunctions caused the wreck is a separate question.

By the time Wegener died, it should have been obvious to most scientists that some sort of theory like continental drift was necessary. Biologists, for instance, recognized that some sort of connections between the continents were required to explain the disjunct distributions of many plants and animal groups across the globe. Nevertheless, they preferred to invoke a fantastic and improbable arrangement of lost continents and land bridges rather than accept the proposal that the current arrangement of continental landmasses and ocean basins might not have been constant throughout the earth's history. The stubbornness on this point seems incredible based on what scientists *did* know about the structure of the earth by the time Wegener published the fourth—and last—edition of his book.

For instance, early oceanographic surveys had indicated that the average depth of the oceans was about two and one-half miles. Without some form of drift, marsupials traveling between Australia and South America would have needed a land bridge something like eight thousand miles long. Such a connection between the two continents—even a fairly narrow one—would have required an almost unimaginable amount of material in order for it to be constructed, but the majority of scientists mired in the comfort of their fixed-earth prejudices had no trouble accepting the idea that such a massive structure could disappear without a trace!

This is not to say that there is no such thing as a land bridge. One important one ran through the Bering Strait, between Russia and Alaska—and it makes perfect sense, for the waters of the strait are fairly shallow and dry land can easily be exposed during ice ages, when sea levels are lower than today. Another exists between North and South America through Panama—although its benefit to migrating animals has been diminished somewhat by the completion of the Panama Canal. The fact remains, however, that most of the land bridges invoked by researchers to explain the distributions of plants and animals were constructs of the mind, not of the earth.

Since the late 1500s, quite a few theories have been hatched invoking large-scale motions of the earth's crust. Why did Wegener's ideas survive while those of others, such as Frank Taylor, fade from memory? Most "drift" theories prior to Wegener invoked fantastic catastrophes, such as Noah's flood, the birth of the moon from the interior of the earth, or the trapping of the moon into earth's orbit. Most were not as ambitious as Wegener's. Taylor seemed to care only about explaining the existence of a belt of mountains near the equator—so what? Wegener sought to explain a host of phenomena with his theory. By treading on turf claimed by specialists in a number of disciplines, I think he offended enough of them to ensure lasting notoriety, which, while probably painful in his lifetime, guaranteed remembrance of his theory after his death. Wegener's predecessors, including Taylor, tended to write a paper or two and move on to some other problem. Wegener never quit working on continental drift, and the fact that he kept fighting to defend the theory until his death kept it in the minds of his colleagues.

Johannes Georgi later said that Wegener would have opposed any attempt to make him a hero or to claim that he never made mistakes. Sure, Wegener did make some mistakes. One was in not seeing himself as a hero. He is to me.

MAURICE EWING also remains controversial. Over the years he has been portrayed as a vehement opponent of continental drift who created such an oppressive environment at Lamont that few dared voice any support of the theory. This is unfair. While I do not think Ewing was ever comfortable with continental drift or plate tectonics, and while he did not contribute many ideas, he more than anyone is responsible for acquiring the data that made such a decisive triumph for plate tectonics possible.

Ewing was a stickler for details, and he had an encyclopedic knowledge of the world's ocean floors. He was reluctant to embrace seafloor spreading because he knew of data that did not fit the hypothesis as originally proposed by Harry Hess and Robert Dietz. He was reluctant to embrace the Vine-Matthews-Morley hypothesis because he could not rule out all alternatives. Ewing, however, encouraged Lamonters to test controversial hypotheses like these. When the hypotheses passed the tests, he made no attempt to suppress the resulting revolution.

WHILE BRUCE HEEZEN was a brilliant and versatile oceanographer, his tremendous success has eclipsed the contributions of Marie Tharp, who for years worked in his shadow. The maps that had such a tremendous impact on how we see the ocean floor would not have been possible without her. Tharp remains relatively unknown today, to some extent a result of her unrelenting efforts to keep Heezen's memory alive in the decades since his death. Tharp's home is in many ways a shrine to three men: her father, her brother, and Heezen, but mostly to Heezen. She remains as devoted to him today as she was throughout their thirty years together as colleagues and possibly more—the extent of their relationship has been a mystery to friends and associates for decades.

Recently Tharp has begun to receive some of the recognition she deserves. The Library of Congress honored her in November 1997 as one of four individuals "who have made major contributions to the field of cartography." The Women's Committee at

Woods Hole Oceanographic Institution presented her with its Women Pioneers in Oceanography Award in March 1999. She was invited to speak at a joint meeting of the California Map Society/Lee Phillips Society in April 2000.

Ironically, Columbia University, at about the same time as the ceremony at the Library of Congress, slighted Tharp. A March 6, 1998, article in the *Columbia University Record* described her as a Columbia cartographer who mapped the Atlantic Ocean floor. Yes, she did that. But the university seems to have forgotten that she mapped the rest of the world's oceans as well.

Fortunately, the memory loss in Manhattan has not affected the Lamont-Doherty Earth Observatory. Tharp was honored along with a number of the people in this story at its fiftieth anniversary celebration in October 1999.

HARRY HESS did not live long past the revolution. He had never taken good care of himself, and the cigarettes and booze caught up with him on August 25, 1969, in Woods Hole, Massachusetts, when he collapsed of a heart attack while chairing a meeting of the Space Science Board. Despite his leadership roles in national and international scientific organizations, despite his rank as rear admiral when he retired from the United States Naval Reserve, despite his status as a giant among geologists, he never took himself too seriously.

"As a geologist who has often guessed wrong," Hess said as he accepted the Geological Society of America's Penrose Medal in 1966, "I deeply appreciate the generosity of the Society in balancing my errors against deductions of mine not yet proven incorrect. I am pleased to come out with a positive balance."

TUZO WILSON retired from the University of Toronto in 1974, but, instead of relaxing, he took over as director of the Ontario Science Centre, at the time the world's largest public museum of science. While there he pioneered "hands-on" exhibits that encouraged visitors to participate in science. He also was instrumental in the development of traveling exhibits that reached remote parts of Ontario. Wilson, always an innovative lecturer, thus made science more approachable and exciting for millions. He retired again in 1983, only to take the position of chancellor of York University, which he held until 1987, when he retired for good. He died on April 15, 1993, at the age of eighty-four.

IN DECEMBER 1975 *Vema* was surveying the Indian Ocean south of Australia when it logged its millionth mile as a research vessel. *Vema* was the first to do so. (Only one other ship has reached that milestone since, in October 1985, and it was another Lamont workhorse, the *Robert D. Conrad.* The *Conrad* was retired four years later.) At the time, *Vema* was working at about fifty degrees south latitude, where nasty weather prevails. No one on board was in much of a mood to celebrate the achievement.

Vema logged another 250,000 miles before Columbia retired it in 1981. During the ship's twenty-eight-year career with Lamont, it had been around the world several times and had sailed in all the world's oceans, once seeking shelter within the flooded volcanic crater on Deception Island, off the coast of the Antarctic Peninsula. Another time *Vema* reached within five hundred miles of the North Pole.

Mike Burke, owner of Windjammer Barefoot Cruises in Miami Beach, Florida, bought the ship in 1981. He rechristened it the *Mandalay.* At the time, the ship looked nothing like the graceful schooner of old. The masts had been mostly removed, the teak decks had been ripped out, and large steel cabins fore and aft had made it resemble a small freighter. Burke set about to fix all that.

Burke had *Mandalay* refit as a barkentine, similar to a schooner but with square-rigged sails on the foremast. Two years later, when the renovations were complete, *Mandalay* began its new career as a cruise ship in the Caribbean. Today it carries as many as seventy-two passengers to locations in the Windward and Leeward Islands and Venezuela. The atmosphere is more relaxed and the accommodations more inviting than during the years when *Vema* was sailed by Lamont, but one can still feel the history that has unfolded on its decks.

NOTES ON REPORTING

During the research for this book, I consulted a number of sources including scientific papers and monographs, published and unpublished memoirs, public and private documents, oral history interviews, and my own interviews with the participants or their associates. The primary sources for each chapter are given below.

Death on a Glacier

The Alfred Wegener quotes came from *Greenland Journey: The Story of Wegener's German Expedition to Greenland in 1930–31 as Told by Members of the Expedition and the Leader's Diary,* edited by Else Wegener, translated from the 7th German edition by Winifred M. Deans (London: Blackie & Son Limited, 1939). The excerpt from Johannes Georgi's letter to his wife came from his 1934 book *Mid-Ice: The Story of the Wegener Expedition to Greenland,* Translated (revised and supplemented by the author) by F. H. Lyon (London: Kegan Paul, Trench, Trubner & Co.).

Radical Notions

The translation of the Abraham Ortelius's passage was taken from James Romm's 1994 paper, "A New Forerunner for Continental Drift" (*Nature* 367:407–408).

Sowing the Wind

The excerpt from Alfred Wegener's 1910 letter to his fiancée, Else Köppen, appeared in Fritz Loewe's 1970 paper, "Alfred Wegener—His Life and Work" (*Australian Meteorological Magazine* 18:177–190).

Reaping the Whirlwind

The translations of Max Semper's and Lucas Fernando Navarro's comments on Wegener's theory were taken from Albert V. Carozzi's 1985 paper, "The Reaction in Continental Europe to Wegener's Theory of Continental Drift" (*Earth Sciences History* 4:122–137). Johannes Georgi's quote, and the excerpt from Wegener's letter to Georgi, came from a chapter by Georgi titled "Memories of Alfred Wegener," which appeared in the 1962 book *Continental Drift,* edited by S. K. Runcorn (New York: Academic Press). All the quotes from the 1926 American Association of Petroleum Geologists' symposium in New York appeared in the proceedings published two years later, edited by W.A.J.M. van Waterschoot Van der Gracht, *Theory of Continental Drift: A Symposium on the Origin and Movement of Land Masses both Inter-Continental and Intra-Continental, as Proposed by Alfred Wegener* (Tulsa, Okla.: American Association of Petroleum Geologists). Tuzo Wilson's quote came from his 1963 paper "Evidence from Islands on the Spreading of Ocean Floors" (*Nature* 197:536–538).

Challenger

Henry Nottidge Moseley's quotes came from his 1879 book, *Notes by a Naturalist on the "Challenger": being an account of various observations made during the voyage of the H.M.S. "Challenger" round the world, in the years 1872–1876, under the commands of Capt. Sir G.S. Nares and Capt. F.T. Thomson* (London: Macmillan and Company).

The Swinger

Vening Meinesz's comments about the *K II* and *K XIII* expeditions came from his book, *Gravity Expeditions at Sea, 1923–1930,* volume I, *The Expeditions, the Computations, and the Results* (Delft, Netherlands: Netherlands Geodetic Commission). The anecdote about Vening Meinesz not lasting twenty-four hours in a sub appeared in A. A. Manten's 1966 article, "In Memoriam: F. A. Vening Meinesz" (*Tectonophysics* 3:369–373). Harry Hammond Hess's recollections were published in his 1957 paper, "The Vening Meinesz Negative Gravity Anomaly Belt of Island Arcs 1926–1956" (*Koninklijk Nederlands Geologisch Mijnbouwkundig Genootschap, Geologische Serie* 18:183–188). The story of Vening Meinesz's visit with Maurice Ewing at Lamont Geological Observatory appeared in the 1966 obituary written by B. J. Collette, "In Memoriam: Professor Dr. Ir. Felix Andries Vening Meinesz, July 20, 1887–August 10, 1966" (*Geologie en Mijnbouw* 45:285–290). J. Lamar Worzel provided the information on Vening Meinesz's role with Netherlands resistance fighters during World War II.

Wildcatter

This chapter was based in part on an unpublished memoir by and discussions with J. Lamar Worzel, on an interview with John Ewing, Maurice Ewing's brother, in 1997, and on several telephone conversations with Betty Ewing, John Ewing's wife, in 2000. The Maurice Ewing quotes were taken from William Wertenbaker's *The Floor of the Sea: Maurice Ewing and the Search to Understand the Earth* (Boston: Little Brown and

Company, 1974). Worzel's quotes came from his contribution to a volume commemorating the fiftieth anniversary of Lamont-Doherty Earth Observatory, "Roots and Rhythms: The Gestation and Birth of Lamont," in *Lamont-Doherty Earth Observatory of Columbia University: Twelve Perspectives on the First Fifty Years, 1949–1999*, edited by Laurence Lippsett (Palisades, N.Y.: Lamont-Doherty Earth Observatory of Columbia University, 1999). The anecdote Sir Edward Bullard related about Ewing, *Atlantis*'s Captain Fred McMurray, and the box of TNT appeared in Bullard's chapter, "Maurice Ewing," which appeared in the 1977 symposium volume *Island Arcs, Deep Sea Trenches and Back-Arc Basins*, edited by Manik Talwani and Walter C. Pitman III (Washington, D.C.: American Geophysical Union).

A Rumor of War—and the Real Thing

Much of the material this chapter was based upon came from an unpublished memoir by and discussions with J. Lamar Worzel. Maurice Ewing's quote was taken from Wertenbaker's *Floor of the Sea*. Worzel's quotes came from his contribution, "Roots and Rhythms," in *Lamont-Doherty Earth Observatory*.

Gunfire and Geology

George Hess, son of Harry Hess, talked to me in 2000 about his father and supplied much of the other material on which this chapter is based, including unpublished letters from Harry Hess to John C. Maxwell in 1944 and 1945 and Harry Hess's diary entries during the United Statess invasion of Iwo Jima in 1945. Harry Hess's remarks regarding his work in Africa and arrival at Princeton were taken from his acceptance speech for the Geological Society of America's Penrose Medal in 1966, published in "Response by Harry Hammond Hess" (*Proceedings of the Geological Society of America* [1966]:85–86).

The Upstarts

This chapter is based in part on J. Lamar Worzel's unpublished memoir, an interview with John Ewing, Maurice Ewing's brother, in 1997, and an interview with William Ryan, also in 1997. The anecdote about the cook quitting on *Atlantis*'s first expedition to the Mid-Atlantic Ridge—including Maurice Ewing's quote about working the ship twenty-four hours a day—came from Susan Schlee's 1978 book, *On Almost Any Wind: The Saga of the Oceanographic Research Vessel Atlantis* (Ithaca, N.Y.: Cornell University Press). The remaining Maurice Ewing quotes, as well as those of David Ericson, came from Wertenbaker's *Floor of the Sea*. Worzel's quote about the founding of Lamont came from his contribution, "Roots and Rhythms," in *Lamont-Doherty Earth Observatory*.

Vema

J. Lamar Worzel's quotes about the acquisition of *Vema*, as well as much additional information, were taken from his unpublished memoir. Captain Louis Kenedy's quotes came from Richard Thruelsen's article "The Incredible Captain Kenedy," which appeared in the January 9, 1954, issue of *The Saturday Evening Post*. Maurice Ewing's account of the purchase of *Vema* came from Wertenbaker's *Floor of the Sea*. Much of the account of the 1954 storm in which Ewing and three others were washed overboard—including the quotes and excerpt from Ewing's letter to his children—also came from *Floor of the Sea*.

The Gully

Marie Tharp provided much of the information for this chapter in a series of interviews in 1997 and in numerous discussions since. The concluding anecdote about Heezen's talk at Princeton was based on my interviews with Tharp and on Walter Sullivan's account in his 1991 book, *Continents in Motion: The New Earth Debate*, 2d ed. (New York: American Institute of Physics). Most of the details came from Sullivan; the exact wording of Harry Hess's concluding statement is from Tharp's version.

Shaken—and Stirred

A portion of this chapter is based on an interview with Dennis E. Hayes in 1999 and one with Fred Vine in 2000. Ewing's quoted answer to Teddy Bullard's question about where Doc kept Lamont's ships was taken from Bullard's memorial to Doc, titled, "William Maurice Ewing: May 12, 1906–May 4, 1974" (*Biographical Memoirs* 51:118–193). Ewing's quip in response to prodding by a Woods Hole Oceanographic Institution colleague about Lamont's rather poor home port facilities was taken from Dennis E. Hayes's contribution, "Ships and Such: Gathering the Flowers of the Sea," in *Lamont-Doherty Earth Observatory*. Henry William Menard's assessment of oceanographic research efforts in the years following World War II was taken from his 1986 memoir, *The Ocean of Truth: A Personal History of Global Tectonics* (Princeton, N.J.: Princeton University Press).

The Paleomagicians

James Heirtzler, Walter Pitman, and Fred Vine provided much of the information upon which this chapter is based during their interviews and discussions with me in 2000 and 2001. Patrick Blackett's quote about paleomagnetism and continental drift was taken from Walter Sullivan's *Continents in Motion*.The story of the rejection of Lawrence Morley's hypothesis comes primarily from William Glen's 1982 book, *The Road to*

Jaramillo: Critical Years of the Revolution in Earth Science (Stanford, Calif.: Stanford University Press).

Revelation

Much of the information about Tuzo Wilson's life and character was obtained from interviews with W. R. Peltier and Fred Vine in 2000. Tuzo Wilson's quip about his enjoyment at disturbing scientists was taken from Peltier's 1994 memorial to him, "John Tuzo Wilson (1908–1993)" (*Eos* 75:609–612). Wilson's quotes about his early career were taken from his 1981 article, "Movements in Earth Science" (*New Scientist* 92:613–616).

Revolution

This chapter is based on interviews with James Heirtzler, Jack Oliver, Xavier Le Pichon, Walter Pitman, Lynn Sykes, and Fred Vine in 2000. Sykes's comment on the significance of the symposium on continental drift at the Goddard Institute of Space Studies in New York in 1966 was taken from his contribution, "Earth-Shaking Events: Seismology, Plate Tectonics, and the Quest for a Comprehensive Nuclear Test Ban Treaty," in *Lamont-Doherty Earth Observatory.*

Alpine Vistas to Abyssal Plains

Much of the information used in this chapter was obtained from interviews with Marie Tharp in 1997, interviews with Billy Glass, Jeff Fox, Xavier Le Pichon, and Bill Ryan in 2000, and an unpublished memoir by J. Lamar Worzel. Tharp's quotes come from my profile of her published in the November/December 1999 issue of the magazine *Mercator's World* (David M. Lawrence, "Mountains under the Sea: Marie Tharp's Maps of the Ocean Floor Shed Light on the Theory of Continental Drift," *Mercator's World* 4[6]:36–43). Some of the information about the controversy over the Second International Oceanographic Congress in Moscow comes from handwritten notes by Heezen describing his side of the dispute. Opdyke's quote was taken from Glen's *Road to Jaramillo.*

Summing Up

John McPhee's synopsis of plate tectonics appeared in his book *Basin and Range* (New York: Farrar, Strauss and Giroux). Rollin Chamberlin's quote from the 1926 American Association of Petroleum Geologists' symposium in New York appeared in the proceedings published two years later, edited by W.A.J.M. van Waterschoot Van der Gracht, *Theory of Continental Drift.* Harry Hess's remarks during his acceptance speech for the Geological Society of America's Penrose Medal in 1966 were published as "Response by Harry Hammond Hess" in the *Proceedings.*

SELECTED BIBLIOGRAPHY

Aldridge, Keith. 1993. John Tuzo Wilson. *The Independent* (London), May 20.

Back at Merchant Marine Academy after Training Trip. 1947. *New York Times*, September 2.

Ballard, Robert D. 1995. *Explorations: My Quest for Adventure and Discovery under the Sea.* New York: Hyperion.

Berton, Pierre. 1988. *The Arctic Grail: The Quest for the North West Passage and the North Pole, 1818–1909.* New York: Viking Press.

Black, George W., Jr. 1979. Frank Bursley Taylor—Forgotten Pioneer of Continental Drift. *Journal of Geological Education* 27:67–70.

Bowin, Carl. 1987. Historical Note on "Evolution Ocean Basins" preprint by H. H. Hess [1906–1969]. *Geology* 15:475–476.

Carozzi, Albert V. 1983. Henrich Wettstein (1880), a Swiss Forerunner of Global Mobilism. *History of Geology* 2(1):41–47.

Columbia Sea Party Is Nearing Bermuda. 1954. *New York Times*, January 15.

Crowley, Thomas J., and Gerald R. North. 1991. *Paleoclimatology.* New York: Oxford University Press.

Dietz, Robert S. 1961. Continent and Ocean Basin Evolution by Spreading of the Sea Floor. *Nature* 190:854–857.

Dott, Robert H., Jr., and Donald R. Prothero. 1994. *Evolution of the Earth.* 5th ed. New York: McGraw-Hill.

Dr. Bruce C. Heezen, 53, Dies; Mapped Ocean Floor. 1977. *New York Times*, June 23.

Dr. Maurice Ewing, Earth Scientist, Dies. 1974. *New York Times*, May 5.

Drake, Ellen T., and Paul D. Komar. 1984. Origin of Impact Craters: Ideas and Experiments of Hooke, Gilbert, and Wegener. *Geology* 12:408–411.

Drum Matthews. 1997. *The Times,* August 12.

Drummond Matthews. 1997. *The Independent* (London), August 1.

Drummond Matthews: Getting to the Bottom of Seafloor Spreading. 1997. *The Guardian* (London), August 14.

Du Toit, Alex. L. 1937. *Our Wandering Continents.* Edinburgh: Oliver and Boyd.

E. F. Huttons to Sell Yacht. 1930. *New York Times,* September 25.

Ellis, Richard. 1996. *Deep Atlantic: Life, Death, and Exploration in the Abyss.* New York: Alfred A. Knopf.

Ericson, David B., and Goesta Wollin. 1967. *The Ever-Changing Sea.* New York: Alfred A. Knopf.

Evolution Linked to Magnetic Field: Extinction of Some Species Is Attributed to Reversals. 1966. *New York Times,* June 2.

Ewing, Floyd F., Jr. 1963. Reverse Migration in West Texas. *Yearbook, West Texas Historical Association* 39:3–17.

Ewing, Maurice. 1938. Marine Gravimetric Methods and Surveys. *Proceedings of the American Philosophical Society* 79:47–70.

Ewing, Scientist, Swept into Sea from Columbia Vessel, Is Saved. 1954. *New York Times,* January 14.

Ewing Tells of Rescue: Says *Vema*'s Pilots Took Him from Fifty-Foot Waves. 1954. *New York Times,* January 16.

Ewing, Maurice, and Bruce Heezen. 1956. Mid-Atlantic Ridge Seismic Belt. *Transactions of the American Geophysical Union* 37:343.

Fast Voyage for *Vema*: Crosses Atlantic in 10 Days, 10 Hours, Breaking Own Record. 1933. *New York Times,* May 24.

Field, R. M. 1945. F. A. Vening Meinesz—a Geodesist's Contribution to Geoscience. *Transactions, American Geophysical Union* 26:183–190.

Fisher, O. 1882. On the Physical Cause of the Ocean Basins. *Nature* 25:243–244.

Flipping the Magnetic Field. 1966. *Time,* June 16.

Fox, Paul J. 1992. Bruce C. Heezen: A Profile. *Oceanus* 34(4):100–107.

Fry, Mervyn C. 1973. Radio's First Voice . . . Canadian! *The Cat's Whisker: Official Voice of the Canadian Vintage Wireless Association* 3(1):2–7.

Glen, William. 1982. *The Road to Jaramillo: Critical Years of the Revolution in Earth Science.* Stanford, Calif.: Stanford University Press.

Greene, Mott T. 1984. Alfred Wegener. *Social Research* 51:739–761.

———. 1998. Alfred Wegener and the Origin of Lunar Craters. *Earth Sciences History* 17(2):111–138.

Heezen, Bruce C. 1960. The Rift in the Ocean Floor. *Scientific American* 203:98–110.

———. 1962. The Deep-Sea Floor. In *Continental Drift,* edited by S. K. Runcorn. New York: Academic Press.

Heirtzler, J. R., G. O. Dickson, E. M. Herron, W. C. Pitman III, and X. Le Pichon. 1968. Marine Magnetic Anomalies, Geomagnetic Field Reversals, and Motions of the Ocean Floor and Continents. *Journal of Geophysical Research* 73:2119–2136.

Heirtzler, James R., Xavier Le Pichon, and J. Gregory Baron. 1966. Magnetic Anomalies over the Reykjanes Ridge. *Deep-Sea Research* 13:427–443.

Hellman, Hal. 1998. *Great Feuds in Science: Ten of the Liveliest Disputes Ever.* New York: John Wiley and Sons.

Hess, H. H. 1962. History of Ocean Basins. In *Petrologic Studies: A Volume to Honor A. F. Buddington,* edited by A.E.J. Engel, H. L. James, and B. F. Leonard. New York: Geological Society of America.

Holmes, Arthur. 1965. *Principles of Physical Geology.* New York: The Ronald Press Company.

The Holy Bible. Containing the Old and New Testaments translated out of the original tongues: and with the former translations diligently compared and revised, by his Majesty's special command. Authorized King James Version. 1985. Grand Rapids, Mich.: Zondervan Bible Publishers.

Isacks, Bryan; Jack Oliver; and Lynn D. Sykes. 1968. Seismology and the New Global Tectonics. *Journal of Geophysical Research* 73:5855–5899.

James, Harold L. 1973. Harry Hammond Hess: May 24, 1906-August 25, 1969. *Biographical Memoirs* 43:108–128.

Lamb, Jamie. 1993. Scientist Got His Energy from Parents. *Vancouver Sun,* April 21.

Laudan, Rachel. 1985. Frank Bursley Taylor's Theory of Continental Drift. *Earth Sciences History* 4:118–121.

Le Grand, H. E. 1988. *Drifting Continents and Shifting Theories.* Cambridge: Cambridge University Press.

Le Pichon, Xavier. 1968. Sea-Floor Spreading and Continental Drift. *Journal of Geophysical Research* 73:3661–3697.

———. 1989. The Birth of Plate Tectonics. *Lamont-Doherty Geological Observatory 1989 Yearbook.* Palisades, N.Y.: Lamont-Doherty Geological Observatory.

Leverett, Frank. 1938. Frank Bursley Taylor. *Science* 88:121–122.

Linklater, Eric. 1972. *The Voyage of the "Challenger."* Garden City, N.Y.: Doubleday and Company.

Lippsett, Laurence. 1999. Introduction to *Lamont-Doherty Earth Observatory of Columbia University: Twelve Perspectives on the First Fifty Years, 1949–1999,* edited by Laurence Lippsett. Palisades, N.Y.: Lamont-Doherty Earth Observatory of Columbia University.

Lockwood, Charles A. 1951. *Sink 'em All: Submarine Warfare in the Pacific.* Toronto: Bantam Books.

Longwell, Chester R. 1946. Presentation of the Penrose Medal to Felix Andries Vening Meinesz. *Proceedings of the Geological Society of America 1945,* 121–127.

Lord, Walter. 1955. *A Night to Remember.* New York: Holt, Rinehart & Winston.

Malone, Thomas F., Edward D. Goldberg, and Walter H. Munk. 1998. Roger Randall Dougan Revelle: March 7, 1909–July 15, 1991. *Biographical Memoirs* 75:288–309.

Maritime Service Buys Sailing Yacht: Luxury Schooner to Be Used as a Training Ship. 1941. *New York Times,* February 10.

Marvin, Ursula B. 1985. The British Reception of Alfred Wegener's Continental Drift Hypothesis. *Earth Sciences History* 4:138–159.

Mason, Ronald G., and Arthur D. Raff. 1961. Magnetic Survey off the West Coast of North America, 32° N. Latitude to 42° N. Latitude. *Bulletin of the Geological Society of America* 72:1259–1266.

Maury, M. F. 1859. *The Physical Geography of the Sea: A New Edition, with Important Additions and Revised Charts by the Author.* London: Sampson Low, Son, and Company.

Maxtone-Graham, John. 1988. *Safe Return Doubtful: The Heroic Age of Polar Exploration.* New York: Charles Scribner's Sons.

McKenzie, D. P., and R. L. Parker. 1967. The North Pacific: An Example of Tectonics on a Sphere. *Nature* 216:1276–1280.

Menard, H. W. 1986. *The Ocean of Truth: A Personal History of Global Tectonics.* Princeton, N.J.: Princeton University Press.

Mirsky, Jeannette. 1970. *To the Arctic! The Story of Northern Exploration from Earliest Times.* Chicago: The University of Chicago Press.

Moores, E. M. 1988. Alpine Serpentinites, Ultramafic Magmas, and Ocean-Basin Evolution: The Ideas of H. H. Hess. *Bulletin of the Geological Society of America* 100:1205–1212.

Morgan, W. Jason. 1968. Rises, Trenches, Great Faults, and Crustal Blocks. *Journal of Geophysical Research* 73:1959–1982.

Morison, Samuel Eliot. 1947. *History of United States Naval Operations in World War II.* Volume I. *The Battle of the Atlantic: September 1939–May 1943.* Boston: Little, Brown and Company.

———. 1948. *History of United States Naval Operations in World War II.* Volume III. *The Rising Sun in the Pacific: 1931–April 1942.* Boston: Little, Brown and Company.

———. 1949. *History of United States Naval Operations in World War II.* Volume IV. *Coral Sea, Midway, and Submarine Actions: May 1942–August 1942.* Boston: Little, Brown and Company.

———. 1950. *History of United States Naval Operations in World War II.* Volume VI. *Breaking the Bismarcks Barrier: 22 July 1942–1 May 1944.* Boston: Little, Brown and Company.

———. 1956. *History of United States Naval Operations in World War II.* Volume X. *The Atlantic Battle Won: May 1943–May 1945.* Boston: Little, Brown and Company.

———. 1958. *History of United States Naval Operations in World War II.* Volume XII. *Leyte: June 1944-January 1945.* Boston: Little, Brown and Company.

———. 1959. *History of United States Naval Operations in World War II.* Volume XIII. *The Liberation of the Philippines: Luzon, Mindanao, the Visayas, 1944–1945.* Boston: Little, Brown and Company.

———. 1960. *History of United States Naval Operations in World War II.* Volume XIV. *Victory in the Pacific: 1945.* Boston: Little, Brown and Company.

New Owner of *Vema* to Cruise to Europe. 1934. *New York Times,* June 3.

The New Yacht Hussar IV: Launched at Copenhagen and Will Be Delivered Early in June. 1923. *New York Times,* March 28.

Obituary: Fritz Loewe. 1974. *Australian Meteorological Magazine* 22:21–23.

Obituary of Tuzo Wilson. 1993. *Daily Telegraph,* April 21.

Office of the Chief of Naval Operations, Naval History Division. 1963. *Dictionary of American Naval Fighting Ships.* Volume II. *C–F.* Washington, D.C.: U.S. Government Printing Office.

Oldroyd, David R. 1996. *Thinking about the Earth: A History of Ideas in Geology.* Cambridge, Mass.: Harvard University Press.

Oliver, Jack. 1996. *Shocks and Rocks: Seismology in the Plate Tectonics Revolution.* Washington, D.C.: American Geophysical Union.

Oliver, Jack, and Bryan Isacks. 1967. Deep Earthquake Zones, Anomalous Structure in the Upper Mantle, and the Lithosphere. *Journal of Geophysical Research* 72:4259–4275.

Opdyke, Neil D. 1985. Reversals of the Earth's Magnetic Field and the Acceptance of Crustal Mobility in North America: A View From the Trenches. *Eos* 66:1177, 1181–1182.

Oreskes, Naomi. 1999. *The Rejection of Continental Drift: Theory and Method in American Earth Science.* New York: Oxford University Press.

Paine, Lincoln P. 1997. *Ships of the World: An Historical Encyclopedia.* Boston: Houghton Mifflin Company.

Phinney, Robert A., editor. 1968. *The History of the Earth's Crust.* Princeton, N.J.: Princeton University Press.

Picard, M. Dane. 1989. Harry Hammond Hess and the Theory of Sea-Floor Spreading. *Journal of Geological Education* 37:346–349.

Raff, Arthur D., and Ronald G. Mason. 1961. Magnetic Survey off the West Coast of North America, 40° N. Latitude to 52° N. Latitude. *Bulletin of the Geological Society of America* 72:1267–1270.

Romer, Alfred Sherwood. 1966. *Vertebrate Paleontology.* 3d ed. Chicago: University of Chicago Press.

Rubey, William W. 1966. Presentation of the 1966 Penrose Medal to Harry Hammond Hess. *Proceedings of the Geological Society of America* (1966):83–85.

Rupke, N. A. 1970. Continental Drift before 1900. *Nature* 227:349–350.

Schlee, Susan. 1973. *The Edge of an Unfamiliar World: A History of Oceanography.* New York: E. P. Dutton & Company.

———. 1978. *On Almost Any Wind: The Saga of the Oceanographic Research Vessel Atlantis.* Ithaca, N.Y.: Cornell University Press.

Schwarzbach, Martin. 1986. *Alfred Wegener: The Father of Continental Drift.* Translated by Carla Love. Madison, Wis.: Science Tech.

Setting Sail on a Training Cruise. 1941. *New York Times,* August 19.

Smith, Dan. 1986. Tuzo Wilson: A Man for All Seasons Flying High after Latest 'Retirement.' *Toronto Star,* August 31.

Spiess, F. 1985. *The "Meteor" Expedition: Scientific Results of the German Atlantic Expedition, 1925–1927.* New Delhi: Amerind Publishing Company.

Spiess, Fred Noel. 1996. Allyn Collins Vine: 1914–1994. *Memorial Tributes* 8:274–279.

Stommel, Henry M. 1994. Columbus O'Donnell Iselin: September 25, 1904–January 5, 1971. *Biographical Memoirs* 64:164–186.

Sullivan, Walter. 1976. A Once Lavish Yacht Is Honored on Millionth Mile of Vital Research. *New York Times,* January 6.

———. 1991. *Continents in Motion: The New Earth Debate.* 2d ed. New York: American Institute of Physics.

———. 1993. John Tuzo Wilson, 84, Is Dead: Early Backer of Continental Drift. *New York Times,* May 30.

Sykes, Lynn R. 1967. Mechanism of Earthquakes and Nature of Faulting on the Mid-Oceanic Ridges. *Journal of Geophysical Research* 72:2131–2153.

Taylor, Frank Bursley. 1910. Bearing of the Tertiary Mountain Belt on the Origin of the Earth's Plan. *Bulletin of the Geological Society of America* 21:179–226.

Texas State Historical Association. 2000. Ewing, William Maurice. In *The Handbook of Texas Online.* Website. http://www.tsha.utexas.edu/handbook/online/articles/view/EE/few3.html.

Tharp, Marie. 1982. Mapping the Ocean Floor—1947 to 1977. In *The Ocean Floor,* edited by R. A. Scrutton and M. Talwani. Chichester, United Kingdom: John Wiley and Sons.

———. 1989. Discovery of the Mid-Ocean Rift System. In *Lamont-Doherty Geological Observatory 1989 Yearbook.* Palisades, N.Y.: Lamont-Doherty Geological Observatory.

———. 1999. Connect the Dots: Mapping the Seafloor and Discovering the Mid-Ocean Ridge. In *Lamont-Doherty Earth Observatory of Columbia University: Twelve Perspectives on the First Fifty Years, 1949–1999,* edited by Laurence Lippsett. Palisades, N.Y.: Lamont-Doherty Earth Observatory of Columbia University.

Theory of Continental Drift. 1927. *AAPG Bulletin* 11:1341–1342.

3-Master Breaks Record in Ocean Crossing; *Vema* Sails to England in 10 Days, 21 Hours. 1932. *New York Times,* May 19.

Totten, Stanley M. 1981. Frank B. Taylor, Plate Tectonics, and Continental Drift. *Journal of Geological Education* 28:212–220.

Tully, Anthony. 1998. Convoy HI-71 and USS *Harder*'s Last Battles. Website. http://www.combinedfleet.com/atully05.htm

Turns Yacht over to U.S.: Owner of *Vema* to Get Only $1 for Luxurious Schooner. 1941. *New York Times*, February 18.

Vacquier, Victor, Arthur D. Raff, and Robert E. Warren. 1961. Horizontal Displacements in the Floor of the Northeastern Pacific Ocean. *Bulletin of the Geological Society of America* 72:1251–1258.

Vema Conquered Rough Seas on Record Trip; Skipper of Schooner on Deck Every Night. 1932. *New York Times*, May 20.

Vema Converted into Training Ship: 3-Masted Auxiliary Schooner Formerly Was One of the Most Luxurious Yachts. 1941. *New York Times*, July 27.

Vening Meinesz, F. A. 1941. *Gravity Expeditions at Sea, 1934–1939*. Volume III. *The Expeditions, the Computations and the Results*. Delft, Netherlands: Netherlands Geodetic Commission.

Vening Meinesz, F. A., and F. E. Wright. 1930. *The Gravity Measuring Cruise of the U.S. Submarine S-21*. Washington, D.C.: U.S. Naval Observatory.

Vine, F. J. 1966. Spreading of the Ocean Floor: New Evidence. *Science* 154:1405–1415.

Vine, F. J., and J. Tuzo Wilson. 1965. Magnetic Anomalies over a Young Oceanic Ridge off Vancouver Island. *Science* 150:485–489.

Vine, Fred. 1993. Obituary: John Tuzo Wilson (1908–1993). *Nature* 363:400.

Vlaar, Nicolaas J. 1989. Vening Meinesz—a Student of the Earth. *Eos* 70(9):129, 134, 140.

Weelden, A. van. 1957. Vening Meinesz, an Integrated Scientist. *Koninklijk Nederlands Geologisch Mijnbouwkundig Genootschap, Geologische serie 1957*, xiii–xvi.

Wegener, Alfred. 1912a. Die Enstehung der Kontinente. *Petermanns Geographische Mitteilungen* 58:185–195, 253–256, 305–309.

———. 1912b. Die Enstehung der Kontinente. *Geologische Rundschau* 3:276–292.

———. 1966. *The Origin of Continents and Oceans*. Translated from the fourth revised German edition by John Biram. New York: Dover Publications.

Wertenbaker, William. 1974a. *The Floor of the Sea: Maurice Ewing and the Search to Understand the Earth*. Boston: Little Brown and Company.

———. 1974b. Profiles: Explorer. I—A Brutal Fact. *The New Yorker* 50 (November 4): 54–118.

———. 1974c. Profiles: Explorer. II—A Landscape That No One Had Imagined. *The New Yorker* 50(November 11): 52–100.

———. 1974d. Profiles: Explorer. III—Some Great Overriding Process. *The New Yorker* 50(November 18): 60–110.

Wilcox, A. A. 1977. Fritz Loewe 1895–1974. *Australian Geographer* 13:306–310.

Wilford, John Noble. 1981. *The Mapmakers: The Story of the Great Pioneers in Cartography from Antiquity to the Space Age.* New York: Random House.

Williams, Frances Leigh. 1963. *Matthew Fontaine Maury: Scientist of the Sea.* New Brunswick, N.J.: Rutgers University Press.

Willis, Bailey. 1910. Principles of Paleogeography. *Science* 31:241–260.

Wilson, J. Tuzo. 1962. Cabot Fault, an Appalachian Equivalent of the San Andreas and Great Glen Faults and Some Implications for Continental Displacement. *Nature* 195:135–138.

———. 1963a. Evidence from Islands on the Spreading of the Ocean Floors. *Nature* 197:536–538.

———. 1963b. Pattern of Uplifted Islands in the Main Ocean Basins. *Science* 139:592–594.

———. 1963c. A Possible Origin of the Hawaiian Islands. *Canadian Journal of Physics* 41:863–870.

———. 1965a. A New Class of Faults and Their Bearing on Continental Drift. *Nature* 207:343–347.

———. 1965b. Submarine Fracture Zones, Aseismic Ridges, and the International Council of Scientific Unions Line: Proposed Western Margin of the East Pacific Ridge. *Nature* 207:907–911.

———. 1965c. Transform Faults, Oceanic Ridges, and Magnetic Anomalies Southwest of Vancouver Island. *Science* 150:482–485.

Worzel, J. Lamar. 1965. *Pendulum Gravity Measurements at Sea, 1936–1959.* New York: John Wiley & Sons.

———. 2000. Tracing the Origins of the Lamont Geological Observatory. *EOS, Transactions of the American Geophysical Union* 81:549–550, 553.

Worzel, J. Lamar, and J. C. Harrison. 1963. Gravity at Sea. In *The Sea: Ideas and Observations on Progress in the Study of the Seas.* Volume 3. *The Earth beneath the Sea: History,* edited by M. N. Hill. New York: John Wiley and Sons.

INDEX

ABOUT THE AUTHOR

DAVID M. LAWRENCE has worked for years as both a scientist and journalist, and holds master's degrees in geography and journalism. He has worked as an independent researcher and research assistant in forest ecology and dendrochronology at a number of institutions—including Lamont-Doherty Earth Observatory—and has served as a reporter, editor, and producer at daily newspapers and on-line sites in Louisiana, New Jersey, and Virginia. Some of his work has appeared in the magazines *Geotimes, Mercator's World,* and *Woods Hole Currents.* He lives in Virginia.

LaVergne, TN USA
24 February 2011
217805LV00001B/1/P